S. Y Richard

The Science of the Sexes

ISBN/EAN: 9783337371722

Printed in Europe, USA, Canada, Australia, Japan

Cover: Foto ©berggeist007 / pixelio.de

More available books at **www.hansebooks.com**

S. Y Richard

The Science of the Sexes

SCIENCE OF THE SEXES

OR

HOW PARENTS MAY CONTROL THE SEX
OF THEIR OFFSPRING

AND

STOCK-RAISERS CONTROL THE SEX OF STOCK

BY

S. Y. RICHARD, A. M.

FIFTH EDITION.

LOUISVILLE:

SPINING & CO., PUBLISHERS.

1879.

CONTENTS.

(iii)

CHAPTER V.

CHAPTER VI.

CHAPTER VII.

CHAPTER VIII.

CHAPTER IX.

CHAPTER X.

CHAPTER XI.

CHAPTER XII.

CHAPTER XIII.

CHAPTER XIV.

CHAPTER XV.

CHAPTER XVI.

CHAPTER XVII.

THE SCIENCE OF THE SEXES.

CHAPTER I.

WHY THIS BOOK WAS WRITTEN.

ABOUT the year 1837 a *quasi* medical work of moderate pretensions fell into the hands of the writer of this, then a young man. He read it with curious and increasing attention. It contained much that was novel and deeply interesting.

In the range of topics it embraced, there was a brief allusion to an immensely important matter—that of CONTROLLING THE SEX OF CHILDREN CONCEIVED IN LAWFUL WEDLOCK. This allusion started a train of thought which has been pursued during the thirty-three years that have since elapsed. Twenty years ago—as far back as 1850—the author persuaded himself that he had fallen upon the true theory of the sexes. It had already been suggested but not satisfactorily proved. As opportunity offered he instituted a series of experiments, and continued his observations with unwearying patience, realizing, in them all, the most abundant and gratifying con-

(7)

firmation of the correctness of this theory. To his own were added the experiments and observations, under even more favorable and remarkable circumstances, of a number of highly intelligent and several scientific gentlemen, more than one of whom was personally and singularly interested in attesting its certainty and truth. Those experiences and processes so greatly impressed the writer that he will incorporate several of them, irregularly and without special acknowledgment, with his own observations and reflections, in some part of these pages.

If this work shall correctly and faithfully record that which, through all these years, has been so perseveringly, and patiently, and hopefully sought out and developed, mankind will have reason to be thankful for this opportunity to learn what might otherwise remain hidden from the masses for some time to come. It is knowledge of a kind that—very quietly and delicately, yet all the more surely—will add materially to the happiness and usefulness of thousands of parents and their families.

THE SEXES.

In all the "wide, wide world" there is not an intelligent or a sympathizing being who has failed to observe the almost desolateness of a home where daughters have no brothers, or sons no sisters. The family circle is shorn of its most beautiful ornament, if there are no sisters to gentleize and polish the natural rudeness of the sons; and if no brother, the daughters seriously, and sometimes sadly, suffer the absence of such an one's whole-hearted companionship and protecting care. We belong to that class— and are thankful that God has planted within us a faith at once so lovely and so grateful—who believe in a *special* not less than in a *general* providence. For the perpetuation of the human species God ordained the sexes, and most wisely and beneficently thus assorts the "olive plants" of the family. At the beginning, "male and female created he them;" so through all generations they are born of one mother, receive one father's blessings, are sheltered by one roof, and fed at a common table. The relation of brother and sister—like the more intimate and indispensable, but scarcely more pure and holy relation of husband and wife—is heavenly in its origin. There is no need, apparently, that every household should embrace children of both sexes; the young of one family flock are not unfrequently all male or all female. And so it might have been from ·the beginning, and still man obey the divine command, "Be fruitful, and multiply, and replenish the earth." There are no parents who so keenly appreciate the value and manliness of sons as those who have daughters only; and none who extol the loveliness and gentleness of daughters so earnestly and unceasingly, as those whose dwellings echo not the music of their sweet voices, nor witness the tenderness of their overflowing sympathy.

THE SEX CAN BE CONTROLLED.

Is it not strange, then, to hear ministers of the Gospel and other good people sighingly exclaim that this thing *can not be helped!* Yet they know that the "wee thing" which is sent in mercy, and becomes a "well-spring of joy," would very often be far more welcome if of a different sex! We tell them they are in error—we are tempted to say they *sin*—in this matter. For just as surely as the patriarch Jacob, in his worldly wisdom [Genesis xxx: 37–42] "took him rods of green poplar, and of the hazel and chestnut tree, and pilled white streaks in them, and made the white appear which was in the rods; and then set these rods before the flocks in the gutters in the watering-troughs when the flocks came to drink; and the flocks conceived before the rods, and brought forth cattle ring-streaked, speckled, and spotted;" and just as surely as he laid these rods before the eyes of the stronger cattle only, and not before the weaker cattle, so that the stronger young cattle were naturally marked as his, while the feebler were Laban's, just so surely, by parity of reasoning, would *similar* worldly wisdom produce strong instead of weakly children, and those almost always *of the sex most desired by the parents.*

If a young man or woman marry a consumptive, no one wonders that their offspring have weak lungs and are short-lived; and so, if those within the interdicted blood-relationship intermarry, this simple and sinful fact most usually explains the cause of any deformity or imperfection, or mental or physical weakness, in their children. If, then, there are natural laws like these which are well known, and whose neglect is painfully visited, why is it strange that there should be a natural law controlling the *sex* of human offspring? Inasmuch as the birth of a male is oftentimes of immense importance, it is not unreasonable that such a law should exist. The highest enjoyments of social and domestic life make its operation necessitous sometimes, and always desirable. The fact that the law itself has been so long *unknown*, does not convince the intelligent mind that there is no such law. There are many results in nature— many gradual but unfailing operations of natural laws—whose processes and causes are yet unexplored. It is but twenty-five years since the electric telegraph flashed its immortal light upon the world; and a century ago the terrific effects and tremendous motive-power of steam were uncalculated, because then undiscovered.

PROGRESS IN KNOWLEDGE.

We have little respect for that person's intelligence and liberality, even though his stock of acquired knowledge be quite considerable, who contents himself with denying that there is a cause—ascertainable, if not in some measure controllable—for *some* of the every-day occurrences in surrounding animated

nature, simply because human progress has not yet
developed it.   A mind thus easily contented would
soon settle the limit beyond which development, material and intellectual, "could no further go."   This
limit once fixed, the very soul of man—so far as
worldly aspirations and encouragements serve to en-
large and ennoble its powers—would be dwarfed and
begin to decline.   Happily, the man of true enterprise is free from such shackles.

" No pent-up Utica contracts his powers;"

more "boundless" than a "continent" is the field
he seeks to explore.   To such a mind and such a
reader—ever willing to learn more of the mysterious,
and ever ready to be convinced of error and taught
the right—regarding all things as "possible with
God," and the sphere of human knowledge as boundless as space itself—to such we address this little
work ; with such it is a pleasure to reason.

THE DESIRABLENESS OF THE KNOWLEDGE.

The desirableness of the knowledge how to *control
the sex of children,* all will concede, excepting,
always, that class who possess and cultivate a blind,
and not an intelligent, faith in the wisdom and goodness of God.   These, if this peculiar knowledge were
offered to them, would decline to use it, lest they
seem " to fight against God."   They regard this sort
of knowledge as purposely withheld from man for all

time to come, simply because it has not been imparted to *them* and is not generally known. Yet they know that God works by means—in this, as in every thing else—in our day. Miracles, in the true sense of that term, have ceased, ages ago. Yet there is much that happens nowadays, and is seen, and done, and learned, that approaches the miraculous. But human reproduction and generation is of every-day life—never miraculous but once. ONE, only, was "conceived of the Holy Ghost," the whole human race by natural means.

CHILD-BEARING THE GREAT LAW OF WOMAN'S NATURE.

At the beginning God established a *rule* for all animate creation: "Be fruitful, and multiply, and replenish the earth."—Genesis i: 28. Man was made male and female, and blessed with fruitfulness and increase. Of the rest of the creatures it would seem that God made many couples, but of man, "*did not He make one?*"—Malachi ii: 15. Angels were not made male and female, for they were not to propagate their kind. And before the fall, before sin brought death into the world to be a never-ceasing check upon the too rapid increase of the human family, there was no necessity for many births or for rapid additions to the population. But the sin in the Garden of Eden brought a penalty with it to each—to the serpent, to the man, and to the woman:

" Unto the woman the Lord God said, I will greatly multiply thy sorrow and thy conception; in sorrow thou shalt bring forth children."—Genesis iii: 16. Thus, as one of the fruits of sin, conceptions were greatly increased, and the pains and pangs of child-birth and abortions, miscarriages and deformed offspring, follow in the train. Of all that bear young, woman is the greatest—indeed, almost the only—sufferer. Her conceptions should be multiplied, and the sorrows of child-bearing become a Scripture-proverb; otherwise death—until the fall, unrevealed among the hidden mysteries of the Almighty—would now destroy the race or prevent the earth from being replenished. And throughout all time—until man shall be utterly destroyed from off the face of the earth—shall woman bring forth " in sorrow," and not with the thrill of joy that Eve felt as she exclaimed, at the birth of her first-born, " I have gotten a man from the Lord."—Genesis iv : 1.

BARRENNESS A REPROACH.

Thus in making man at the beginning male and female, God established child-bearing as the *rule*, and ordained it as the law of woman's nature. Barrenness is the *exception* to the rule—a violation of, or exemption from, this law of nature. The women of the Bible regarded it as a " reproach " if they did not bear children, and in bitterness* of heart complained of their husbands, and prayed God to remove from them this reproach. Rachel, the beautiful wife of the patriarch Jacob, envied her sister Leah because she saw her "endowed with a good dowry"—six sons—while she herself remained unfruitful. But at last "God remembered Rachel, and hearkened to her, and opened her womb, and she conceived and bare a son, and said God hath taken away my reproach."—Genesis xxx: 23.

RACHEL'S WISDOM.

In her exultation at the joyful event, Rachel exhibited the singular wisdom that seems most generally to be denied to mothers at the present day: "And she called his name Joseph," [which means *adding*] "and said, The Lord shall add to me *another son!*" Now that her barrenness was cured, her womb opened, and the natural law of child-bearing thus established in her person, she confidently and not unreasonably looked forward to having *another* child. But *that child should be a* SON! Why should she say that? How could she know it? Was Rachel "also among the prophets?" Commentators do not even suggest, much less claim, that she was! They content themselves with saying that "her enthusiastic looking forward may be regarded as the language of her inordinate desire for another son (she scarcely knows how to be thankful for one, unless sure of another), or, it is the expression of her faith (she takes this mercy as an earnest of further mercy)."

It is not strange that commentators should be puzzled to comprehend the source of Rachel's knowledge, any more than the depth of her feelings. Jacob had two wives; Leah was forced upon him, and was not his favorite; he loved Rachel more, serving fourteen years for her, "and they seemed unto him but a few days for the love he had to her." But God, for His own wise purposes, had shut up Rachel's womb, whereas Leah bore her husband six sons, and the two handmaids of the two wives each bore him two sons—in all, *ten sons!* Why were these ten children in succession, born of three women, *all* sons? Were they the result of accident—of ten successive accidents? Had the parents, at conception, no control over the sex of their offspring? Were those women incapable of bearing, or Jacob of begetting, daughters? The *eleventh* child—Leah's *seventh* and *last*—was a daughter, Jacob's *only* daughter!

Envy and jealousy, intensified by the feeling that the "reproach" of women was upon her, tormented the breast of the beautiful Rachel. In bitterness of spirit she demanded of her husband, "Give me children, or else I die." Jacob angrily reproved her, "Am I in God's stead, who hath withheld from thee the fruit of the womb?"—Genesis xxx: 2. This is proof incontrovertible that child-bearing was ordained as the general or universal law of woman's nature. The womb was made to bear its fruit; marriage is to produce the fruit; if withheld, it is God withholds it. In case of barrenness there is some serious interruption of the most usual course, some violation of the general law—always resolvable back to God, although He frequently allows human means to overcome the interruptions and restore the natural operation of the law of child-bearing.

METHODS TO CURE BARRENNESS.

Another extraordinary fact here : Rachel, in her anxiety to bear children, resorted to unusual means. She not only claimed and enjoyed much the larger portion of her husband's attention, but she ate mandrakes (a kind of love-apples) to exhilarate her spirits and excite to love.—Genesis xxx : 14, 15, 16. If, then, she knew of and positively employed artificial means to stimulate to venery, assuredly she would control the sex of her children and bear only boys, if within her knowledge and power. There need not be a doubt that this power to control the sex was known to these four women, and in all these cases was exercised. God works by natural means— seldom by miracle. He " makes the wrath of man to praise Him." While the will of man is free—in his tastes, his desires, his ambition, his covetousness almost untrammeled—God, nevertheless, overrules and disposes. The promised Seed was to come out of Jacob's loins. If Rachel bear no children, her " reproach " would be too great to be endured ; if she have daughters, but no sons, her offspring must assuredly lose the honoring inheritance ; the sons of

2

Bilhah, *her* handmaid, could not be in the favored
line of succession, because Bilhah was a "*bond-
woman.*" Rachel *must* bear *sons,* or painfully realize
that the promised Seed would come through one of
Leah's sons. No wonder, then, that she should bear
sons *only,* if her knowledge and power should equal
her desire.

FIRST DEATH IN CHILD-BED.

Rachel's history, in bearing another son, as she so
confidently said she would—the result of human
knowledge, as we verily believe, and without relation
to and independent of the gift of prophecy—presents
another sad and notable fact: The first recorded
case of the pains and perils of child-bearing, which
was thus introduced by, or was the effect of, *sin!*
She travailed and "had hard labor," and died just
after naming the new-born son.

FURTHER AS TO CONTROLLING THE SEX.

But this first known instance of the mother's death
from child-birth is strong confirmation of the view
that the control of the sexes was, at that early day,
within the knowledge of those who were wise in such
things. The sacred writer tells us, "It came to
pass, when Rachel was in hard labor, that the midwife said unto her, Fear not, thou shalt have this
son also."—Genesis xxxv: 17. The mother, at the
birth of her first-born, had perpetuated the fact of

her knowledge of this mystery in the very *name* she gave him. And now the midwife, while thereby attesting her own skill and knowledge of these matters, encourages the dying mother to bear patiently her pangs, because another *son* was about to be born of her. Did the midwife obtain her knowledge from Rachel, or had she previously instructed Rachel in the mystery of *son*-bearing? The two passages compared leave no room to doubt that both women were *au fait* on this point of physiological learning or experience. As all of Jacob's children, except the youngest, were natives of Mesopotamia, whereas Benjamin, the youngest, was born near Bethlehem-Judah, full five hundred miles distant, it is not likely that the same midwife attended Rachel in her two confinements. And the tone and language of encouragement that the midwife used just before the last birth was of that positive and decided character seldom used, except by those who *know* whereof they speak. It is quite probable that the same valuable knowledge which enabled Rachel to control the sex of her first-born, was not confined to the women of her native land, but was just as much in the line of medical or midwife skill and science in the land of her adoption. It certainly was of more practical importance in later years among the Jews, whose tribes came out of Jacob's loins

CHAPTER II.

WOMAN'S LOVE OF OFFSPRING.

WHILE thinking upon this tender and absorbing element of female nature, our eyes fell upon a sketch written many years ago, and to illustrate a different connection, which so forcibly and so happily expresses what we wished to say, that we embody it here. None but a husband and father could realize such beautiful feelings, or so touchingly portray them:

"Woman is by nature a producer, former, educator of her race. She is instinct with the desire of offspring, which nothing else can satisfy. Her soul is silently, but ceaselessly on fire, with love of progeny. The perils that attend on pregnancy and parturition, sometimes occupy her thoughts; the joys of offspring always.

> " ' Man's love is of man's life a part,
> 'T is woman's sole existence.'

"Her form, her make, her organization, her thoughts and feelings, are expressly constituted, all for offspring. The eye is not more evidently formed for seeing, the hand for holding, and the feet for walking, than is a woman formed for offspring. The instinct, the inwrought desire of woman is for offspring. She is constructed outwardly for this very purpose. Her abdomen and hips are large for the reception and gestation of her offspring ; her lap is ample for its couch and resting-place ; her bosom fitted for its nouriture and fondling ; her limbs and person soft and flexible, to make a gentle, yielding, easily compressible nurse and playmate. Her hands are delicate and exquisitely formed for gently hand-ling tender beings. Her feet are small, her legs constructed to take tiny steps, so suitable and requi-site for those who have the office of accompanying infant locomotions.

" One thing is most remarkable, and yet it seems to have escaped the observation of philosophers and physiologists. The beauty of the woman, both in form and feature, seems to have no adequate use, unless it is a constant object of attention to her wor-shiping offspring. Then the true use of woman's beauty is encircled with a glory, which its delight for man alone would never give. When we con-sider woman's beauty, like the stars of heaven, or flowers of earth, an object of unfailing, never weary-ing joy to children, our estimation of its worth, and its Divine Bestower's goodness, are raised beyond the highest and most pure conception of mere woman worship. Where beauty terminates, as it originates, in goodness, 't is divine.

HER SYSTEM ADAPTED TO CHILD-BEARING.

"The bosom, face, and hair of woman are so much more soft and winning than they are in man, that children are instinctively induced to seek their comfort and enjoyment in her presence. The power to please is always grateful, and nothing can afford a human being purer, richer, more refined and satisfactory enjoyment, than the power of making children happy. Possession of a faculty or power implies, of course, delight in exercise, and the existence of the object requisite to its enjoyment. The highest faculty with which we are endowed, is that of being able to produce, or to create, the objects necessary to our happiness. Such is the privilege of woman. Every development of her mysterious organization is for producing and sustaining offspring. What is the womb? A pear-shaped organ, with a cavity which opens to receive the embryotic seed of a new being, and then instinctively closes and seals itself up, in order that it may incorporate the germ with a miraculous ovum, and nourish and develop it into a fœtus. What are the ovaries, or egg-beds, but two organs, which supply, and periodically send off, the ova or the eggs which steadily and surely seek for impregnation. The satisfaction of the womb is in receiving and retaining. To lose, is just as miserable for the womb, as for the hand or head. It is as impossible for loss to be con-

verted into gain, as for miscarriage to be turned to
happiness for woman!

"The blank unsatisfiedness of the barren womb,
has been proverbial in every age. From the days
of Rachel, who exclaimed, with exquisite, pathetic
longing, '*Give me children or I die*,' to the time
of Solomon, who, in his universal observation of
mankind, has recorded this intense desire, which
says, '*Give, give*,' and never can be satisfied, the
constant testimony of the Scriptures is to the happiness of offspring, and to the wretchedness of sterility
and miscarriage. Nor is there any change in the
preceding or succeeding parts of Scripture. There
is but once a woe denounced on offspring, and a
blessedness pronounced on barrenness, and that was
by the Saviour, in his beautiful lament for doomed
and desolate Judah and Jerusalem. Pity and love
to miserable woman, reversed the blessing and the
curse for once—but not reversed her nature. The
sun is not more native to produce than is the womb
of woman.

"The function of the womb, untended or perverted, is as annoying to a woman as a faculty of
the mind, when left uncultivated. It can not be
completely dormant. It necessarily influences all
the other faculties and functions of the woman.
The faculties of observation and constructiveness, if
not attended to, may be unnoticed and overlooked,
because they are not normally in action; yet will
they manifest themselves irregularly, by the prying,

meddling disposition of the person who possesses them in their abnormal state. The functions of the womb affects the woman in the same manner. If it be rightly tended, whether in an active or passive state, the character is softened, elevated and refined. If it be rudely treated, neglected, or perverted, it gives a roughness, coarseness, and ferocity to woman, almost unsexing her.

" The largeness of the hips and abdomen in woman, imply a prëarranged capacity for bearing children, and the well-known pleasure which a woman feels, when conscious of her pregnancy, and that it adds unto her interest and beauty, are large additions to our argument, that love of offspring is not only natural, but a strong necessity of her being.

" Perhaps the strongest portion of the argument remains to be adduced : The function of the breasts. The beauty of these parts of woman's structure, we have already briefly treated. Give us the breasts, with their rich function of lactation, and we have all the previous functions and performances required for offspring, inevitably guaranteed by all the laws of nature and of Providence. Prominent and commanding must be the desire, the love of offspring. The love of life, the appetite for food, the keen enjoyment of the senses, can not surpass the keenness of desire for, nor the strength of love to, offspring."

ILLUSTRATIONS FROM SHAKSPEARE.

" A few quotations may be made, confirmatory of our theory, with good, and, certainly, agreeable effect. The character of Rosalind, in 'As You Like It,' is one of Shakspeare's highest feminine creations. She is a tall and graceful nervo-sanguine beauty, vivid in her imagination, abundant in the flashes of keen, caustic, but unwounding wit, sunny as summer in her exquisite affections, and every thought and feeling deeply dyed with womanhood; but beautifully, innocently, yet not ignorantly, chaste and pure. The function of her womb diffuses over her a rich and fascinating mellow, moral feeling, which charms and chains admiring and transported man, and lights up woman's fancy brilliantly and elegantly, displaying it in all the glory of a tropical profusion. After the wrestling scene, when young Orlando had excited in her heart, for the first time, the elegant, subduing passion of pure love, she sighs her feelings forth to her well-trusted, sympathizing cousin, Celia. Rosalind's father being now in banishment, Celia, with admirable woman's tact, asks if all this is for her father, and elicits the reply, which we *italicise* as our quotation. We give that portion of the scene where it occurs:

" 'CELIA.—Why, cousin; why Rosalind; cupid have mercy! Not a word?

" 'ROSALIND.—Not one to throw at a dog.

"' CEL.—No, thy words are too precious to be cast away upon curs, throw some of them at me; come, lame me with reasons.

"' Ros.—Then there were two cousins laid up; when the one should be lamed with reasons, and the other mad without any.

"' CEL.—But is all this for your father?

"' Ros.—No, some of it *for my child's father!*'

"Nothing could be more natural, more elegant, and exquisitely feminine. She traces love to its appropriate and desired results, with one of those fine strokes, which woman only has the power to give. She speaks the thought and feeling of her heart—love and fruition.

"In the 'Merchant of Venice,' where Bassanio has chosen the right casket and the majestical but exquisitely simple Portia, dedicates herself and fortune to him, in language which for gentleness and tenderness of sentiment, and elegance of expression, has no parallel—Gratiano and Nerissa confess their love, and the two charmed pairs, betrothed, are now proceeding to their marriage. The merry soul of Gratiano, in the presence of the queenly Portia, and her accomplished maid, Nerissa, fired with the joys of expectation, speaks what he knows will touch the golden chord of feeling in the bosom of them both. Beholding his fair charmer, he exclaims, '*We'll play with them; the first boy, for a thousand ducats.*' Perhaps there is no higher aspiration of a wife, than that her first-born may be a cherub boy. The feeling of delight which thrills the soul of woman, when she contemplates, with reasonable expectation, that

she will bring forth a beauteous image of the being
whom she loves, surpasses all the loftiest emotions
of her love to man. Creation is her glory—offspring is the perfection of her function."

CATHARINE, QUEEN OF HENRY VIII.

"Take one more well selected case from the picture of that paragon of virtue and true beauty of the
soul, as wife and mother, Catharine, queen of Henry
VIII. In that majestical and marvelously moving
pleading, which she makes before the king, presiding
o'er the court, assembled for the purpose of divorcing
her—a pleading, which, for shape, and course, and
argument, and pathetic power, a hundred cardinals
and proctors might in vain essay to compass—she
has this exquisitely apposite, most delicate, and
charming passage:

> "'Sir, call to mind,
> That I have been your *wife* in this obedience
> Upward of twenty years, and *have been blest*
> *With many children by you.*'

"She first appeals to his known strong propensity
for the married life, reminding him, that in the
quality of wife, she had supplied his wants, obe-
dient to his will, for twenty years. Had she staid
here, he might have felt the pain of obligation, a
feeling most inimical to her present cause. She
wisely, beautifully, puts the sense of obligation on

herself: '*I have been blest with many children by you.*'

"The poet shows his master knowledge of the human mind, by this acute perception of the feelings of a pure, high-minded, virtuous wife; lofty in honor, yet a saint in meekness and humility. Had the king only, been the court to which she appealed, she would have gained her suit—for when she left he straight pronounced her eulogy, as fondly as a lover, dwelling on her enchanting qualities as a wife. A commoner poet would have made the queen bring in the king a debtor to her for his children. A commoner woman than Queen Catharine would, inevitably, have so done, and most assuredly have missed her mark."

ILLUSTRATIONS FROM MILTON.

"Of all men not included in the Scripture category of 'inspired,' Milton appears the loftiest, the purest, and the most sublime of mortals. He was the most profound of scholars, a master of the sciences of mind and morals, most thorough in his knowledge of mankind, a mighty statesman, a most comprehensive and acute philosopher, and one of the most sage and grave of theologians. How does he draw the nature and the character of woman? In that divine relation, which he put into the mouth of Eve, recounting to her consort, Adam, her waking up to consciousness of life and being, she tells him that she heard a voice, which said:

" ' But follow me,
And I will lead thee where no shadow stays
Thy coming, and thy soft embraces : he
Whose image thou art, *him thou shalt enjoy,*
Inseparably thine : to him shalt bear
Multitudes like thyself, and thence be
Call'd mother of human race.'

" In this passage, the Deity gives her the promise of enjoyment of her husband, and a multitude of off-spring. This promise, uttered in the ears of modern, fashionably-educated, and perverted woman, would sound more like a curse.

" Again, in that celestial adoration which the first pair offer ere they go to rest, they say :

" ' Happy in our mutual help,
And *mutual love, the crown of all our bliss—*
For thou hast promised from us two, a race
To fill the Earth.'

" The poet himself, in speaking on this subject, says :

" ' *Hail, wedded love, mysterious law, true source*
Of human offspring ! * * * By *thee,*
Founded in reason, loyal, just, and pure,
Relations dear, and all the charities,
Of father, son, and brother, first were known,
Perpetual fountain of domestic sweets.'

" Once more, dilating on this subject, he exclaims :

'*Our Maker bids increase*—who bids abstain ? ' "

ILLUSTRATIONS FROM THE BIBLE.

" In that sublimely brief, and yet elaborate relation of the origin of our race, recorded in the first
of Genesis, the Deity is represented as addressing the
first pair thus: ' *Be fruitful and multiply, and replenish
the earth.*' Here, in the beginning of the human race,
the divine command showed the great end and purpose of its creation.

" The same permission and command, as that
which the first pair received from God, are represented to have been repeated unto Noah, the second
great progenitor of the race, in the benignancy of
the Divine complacence. ' *And God blessed Noah
and his sons,* and said, *Be fruitful and multiply.*' He
repeats his blessed wishes for their happiness, and
makes a covenant of peace and plenty with them
and their offspring, as an inducement for them to
be fruitful.

" The history of Abraham and the Patriarchs is
replete with instances of blessings promised in the
shape of offspring. To Abraham and Sarah, the
great object of their lives, and promise of their God,
was offspring; and when, from age, they deemed
themselves incapable of produce, the supernatural
promise of a son inspired their expectations, and renewed their youthful feelings, so that they laughed
to think that they should yet enjoy the pleasure of
offspring, which had been hitherto so constantly denied them.

" The beautifully simple, yet romantic story of Rebekah, which throws into the shade a thousand gilded scenes of modern novelists, and nowhere has its equal for simplicity and purity, is especially remarkable for the parting blessing of her mother, and her brother, when she left with Abraham's servant to be Isaac's wife: ' *Be thou the mother of thousands of millions.*'

" The strife of rivalry between the sister wives of Jacob for the precedence of offspring, is one of the most artless, interesting stories ever told. The natural simplicity and truthfulness with which the wants and wishes of the rival wives is told, is quite refreshing in these days of prudish purity, but practical indecency. The husband, Jacob, was a prize which each contended for, expressly on the plea of adding to her offspring.

" The giving of their maids to him as wives, for the same purpose, was but carrying out the feeling so intensely manifested in the sisters. What language could express more vividly their feeling, than this exquisite relation ? It was the business of their lives, the burden of their prayers. The exclamation of that queen of nature's women, Rachel, is the crowning passage of the whole : ' *Give me children, or I die.*'

" When Moses brought the Israelites out of Egypt through the wilderness, to the borders of the land of Canaan, before he left them, he told them of the blessings which the Lord had promised them—and foremost, with much emphasis of expression, do we find the joys of offspring : ' *He will love thee, and bless thee, and multiply thee. He will also bless the fruit of thy womb.'* ' *Thou shalt be blessed above all people: there shall not be male or female among you barren!'* This is repeatedly promised, as the greatest of all earthly blessings. We have seen how offspring was coveted by women; we here see how it was cared for, and provided, by the God of Israel."

THE STORY OF HANNAH.

" The touching story of Hannah, Samuel's mother, one of the two wives of Elkanah, in the first book of Samuel, is remarkably expressive of this strong desire for offspring. It is stated that ' *it made her fret because the Lord had shut up her womb—therefore she wept and did not eat ;'* although her husband appealed to her most tenderly, saying, '*Hannah, why weepest thou, and why eatest thou not, and why is thy heart grieved? Am not I better to thee than ten sons?'* She went into the temple of the Lord, 'in bitterness of soul, and prayed unto the Lord, and wept sore; and she vowed a vow unto the Lord, and said, O, Lord of Hosts, *if thou wilt indeed look on the affliction of thy handmaid, and remember me, and not forget thine handmaid, but wilt give unto thine handmaid a man-child,* then I will give him unto the Lord all the days of his life.' And when Eli had mistaken her ejaculations of prayer for the mutterings of drink, she says : '*I am a woman of a sorrowful spirit. I have poured out my soul before God, for out of the*

abundance of my complaint and grief have I spoken.'
Could language be more eloquent, more pungent,
or more moving than the words and prayer of
Hannah? How inexpressibly desiring is the love
for offspring? And when her wish is gratified, how
elegant, how lofty, how sublime her psalm of praises
and thanksgiving! Among the beautiful expressions of her soul, for her long-prayed-for boon, we
find: '*My heart rejoiceth in the Lord, my horn is exalted. I rejoice in thy salvation; they that were hungry
ceased, so that the barren hath borne seven.'* She com-
pares her longing for offspring to the feeling of
hunger, and then employs the Hebrew figure for
completeness, satisfaction, fullness — the number
seven—to set forth the rapture of her high thanksgiving."

THE SWEET SINGER OF ISRAEL.

"We turn now to the Psalms for some of the most
delicate and beautiful quotations on this subject.
The hundred and twenty-seventh and eighth contain some passages of singular felicity for our purpose: '*So children are an heritage of the Lord, and
the fruit of the womb is his reward. As arrows in the
hand of a mighty man, so are the children of youth.
Happy is the man that hath his quiver full of them.'*
'Blessed is every one that feareth the Lord. *Thy wife
shall be as a fruitful vine upon the sides of thy house
—thy children like to olive-plants around thy table.
Behold, thus shall the man be blessed that feareth the*

3

Lord. *Yea, thou shalt see thy children's children.'*
Here offspring is spoken of as a reward of God.
Children are said to be a rich estate—a valuable
possession and protection. A prolific wife is likened
to a fruitful vine. The beauty and the value of the
children are compared to those of olive plants, and
to see your children's children the perfection of
human happiness. The author of these Psalms must
have seen much of human life under its best aspects,
and his philosophy and morality are only equaled
by his poetry.

"The most sublime of dramas, the poetic book of
Job—the loftiest production of the mind of man—
which, for acquaintance with the principles of human
nature, and profusion of refined, majestic, and terrific
imagery has no equal, even in the Scriptures, in recounting the consoling termination of the Patriarch's
trials, gives a detailed account of his prosperity,
which, in the Oriental style, has a magnificent pro-
fusiveness. The last part of this picture of prosperity is: *'He had also seven sons and three daughters.'* This was the crown of all his riches and his
bliss."

THE PROPHET ISAIAH.

"In quoting from the beauties of the Carmel
heights or Succoth valleys of the majestied Isaiah,
we pass to that well known, often quoted, passage:
'Can a woman forget her suckling child, that she
should not have compassion on the son of her

womb?' The following passage is a strong and beautiful one for us: '*Sing, O barren, thou that didst not bear; break forth into singing and cry aloud, that thou didst not travail with child, for more are the children of the desolate than of the married wife.*' Here the intensity of the desire for offspring is accounted so exceeding strong that a woman who has hitherto been barren, when she comes to bear is called on, even in the pangs of labor, to rejoice.

"One of the most beautiful and moving adaptations of this strength of feeling in the woman, will be found in a prophetic passage relating the sufferings and enjoyments of the Saviour, in the work of expiation: '*When thou shalt make his soul an offering for sin, he shall see his seed; he shall prolong his days; he shall see of the travail of his soul, and shall be satisfied.*' This passage, which is so tenderly and so intensely moving, would be a useless waste of idle words if the strong feeling, which we are contending for, were not a principle of human nature.

"Another passage of most exquisite affection, where God consoles his people, is based entirely on this principle: '*Fear not, thou shalt not be ashamed, neither be thou confounded, for thou shalt not be put to shame. For thou shalt forget the shame of thy youth, and shalt not remember the reproach of thy widowhood any more. For the Lord hath called thee as a woman, forsaken and grieved in spirit—as a wife of youth, when thou wast refused.*' For a woman to expect marriage and offspring, and to be disappointed, is perhaps the most mortifying and humiliating occurrence of her being, more especially if it be on account of her supposed incapability of, or unfitness for, the office of production. Except this feeling were a universal one, what wisdom would there be in an allusion to it in a national address?"

THE PROPHET JEREMIAH.

"When the children of Israel were carried away captives to Babylon, the prophet Jeremiah was commanded by God to send them a comforting message, and he wrote to them thus: '*Build ye houses and dwell in them. Plant ye gardens and eat the fruit of them. Take ye wives and beget sons and daughters. Take wives for your sons, and give your daughters to husbands, that they may bear sons and daughters, that ye may be increased and not diminished, and seek the peace of the city.*' Can any thing be plainer teaching than this—that in raising up and cultivating families, the peace and prosperity of the world is secured? When men and women are busily and happily employed at home, the public peace is sure.

THE SHUNAMMITE WOMAN.

"Who that is familiar with the life of Elisha, the prophet, remembers not the beautiful story of the Shunammite and her son! The prophet had been entertained by her for some considerable period, and was thinking what to do for her in return. He told his servant to say to her: 'Behold, thou hast been careful for us with all this care, what is to be done for thee? Wilt thou be spoken for to the king or to the Captain of the Host? And she answered, 'I dwell among my own people.' And Elisha said, 'What, then, is to be done for her? And Gehazi said, '*Verily, she hath no child, and her husband is old! And Elisha said, Call her. And Gehazi called her, and she stood in the door. And Elisha said, About this season, according to the time of life, thou shalt embrace a son. And she said, Nay, my lord, thou man of God, do not lie unto thine handmaid.*' The news was too delightful to be credited."

NEW TESTAMENT ILLUSTRATIONS.

"There might be much more gathered from the Old Testament, but we must close with this relation, and very briefly introduce a few from the New Testament. The first that we shall quote is that most interesting natural relation of the cousins Mary and Elizabeth, as given in the second chapter of the Evangelist Luke. Elizabeth had been a barren woman, and now was pregnant. Inspired with love and gratitude, she broke forth in a song of praise, beautiful and majestic. Among the exquisite expressions which she uses there are these: 'My soul doth magnify the Lord, my spirit hath rejoiced in God, my Saviour. *For he hath regarded the low estate of his handmaiden, and he that is mighty hath done great things for me.* He hath filled the hungry with good things.' Remembering that Elizabeth was now advanced in years, how strong the feeling must have been which made her joy in that which most would deem a peril at her time of life! The domestic feast was much enriched by social thanks. '*She brought forth a son, and her neighbors and her cousins heard how the Lord had showed great mercy on her, and they rejoiced with her.*' It is exceedingly to be regretted that in modern times few would be found to join the neighbors and the cousins in such rejoicing.

"Jesus, who knew so perfectly the thoughts and feelings of both man and woman, said: '*A woman when she is in travail, hath sorrow, because her hour is come, but as soon as she is delivered of the child she remembereth no more the anguish, for joy that a man is born into the world.*' Now this is said of woman. The very act of giving a child birth is joy to her. Could it be otherwise, unless malignancy, instead of the Divine benignancy, had favored her?"

THE APOSTLE PAUL.

"The Apostle Paul, who has been unadvisedly accused of favoring monkish celibacy, and therefore oftentimes far from a favorite with women, in his epistle of advice to Timothy, respecting women, is so very plain and absolute that no one can mistake him: '*I will* (I desire), *therefore, that the younger women marry and bear children.'* Will any man say after this that Paul did not regard the feelings and desires of women?

"Thus we find that Sacred Writ is all confirmatory of the principle that woman was designed for love and offspring."

CHAPTER III.

STERILITY.

THE causes of sterility are, disease of the ovaries, closing of the Fallopian tubes, a diseased uterus, or the displacement of the womb. This disease has been diagnosed by many eminent physicians. We can give no better idea of it than by quoting from a few of them.

An eminent French physician, and author of a popular work on physiology, says in his chapter on Sterility:

" ' Be ye fruitful and multiply,' is a command which should be cheerfully obeyed by the children of men ; and, in fact, it would seem to be the principal end of man's earthly existence, and so imperative is the voice of nature in the matter, that the universal mind of the human race is more concentrated on the feeling that leads to the consummation desired, than on all the other attributes of being put together. No two men are exactly agreed on any other subject, and no two differ on this. All are united in the desire which finds its accomplishment in the reproduction of their kind.

"Wherein lies the mystery of this? There is nothing more certain than that it is one of the first duties of the human race to increase and multiply; and the man who leaves the world without having obeyed the injunction, can scarcely be said to have fulfilled the great end of his existence. 'But,' perhaps replies the reader, 'many men are so constituted that they can not produce offspring?' This I deny; for all men who are not evidently of monstrous conformation, or who have not been seriously injured by artificial means, are equal to the task of reproduction; indeed, without the parts and means necessary for reproduction, he could scarcely exist at all, and would be no more a human being than if he were deficient of heart and brains. Such things, they tell us, have been, but I have never seen any proof of it. I am also convinced that there is no such thing as natural barrenness in *natural* women; and that the causes which are supposed to render them so, can, in ninety-nine cases out of a hundred, be removed.

" However, it can not be denied that a vast number of married persons are unblest with offspring, whose exertions are undoubted, and who would give much, were it otherwise."

CAUSES OF UNFRUITFUL MARRIAGES.

" The causes of unfruitful marriages are numerous. One is, the mutual coldness of the parties; another, the mutual intensity of their desires; a third, unfitness in consequence of the difference of their physical construction for sexual intercourse; besides which, may be mentioned · leucorrhœa, disgust, timidity, extreme indulgence, lowness of spirits, irregular menstruation, etc. Also, the obliteration of the vaginal canal, or absence of the ovaries or Fallopian tube in the female; but these latter are of such rare occurrence, that the parties so afflicted may be put down as vagaries of human nature, and, therefore, monsters; and women so situated, if they know their afflictions, are altogether unfit for the duties of married life, and are guilty of a serious offense in smuggling themselves within its pale. However, one female in every million is not thus circumstanced, and, consequently, one out of every million is not of necessity barren."

UNFITNESS FOR INTERCOURSE. ●

"I said unfitness for intercourse is one cause of unfruitfulness. It is, however, a rare one; for young married persons, probably in the ratio of five hundred to one, become physically adapted to each other, even though there should be some seeming barriers at the commencement. Nevertheless, it frequently happens that a couple will have no offspring; yet, being divorced, and forming other connections, both will have children, which indicates an unfitness for intercourse in the first instance.

Thus it was with Napoleon Bonaparte and Josephine, who, though unfruitful in connection, gave proof that the deficiency did not lie in either, but only in their relationship to each other. I do not think, however, that this marriage was unfruitful from the cause indicated, but that both were too highly intellectual to produce the excitement necessary to the end desired; or rather, that the coldness of each, as to amorous pleasures, repulsed the other. Therefore, it is probable that proper stimulants would have excited a mutual warmth of feeling, and given an heir to the throne of France. Baron Larry expressed the same opinion; and I have it from good authority that the Emperor felt this, and resorted to borax, marjorum, and even Verrey's Tincture of Magnanimity, though without effect; which induced him to repudiate a wife whom he dearly loved, and marry one of stronger desire, to counteract his own absence of animal feeling."

TOO GREAT ARDOR.

" A couple of full habits and strong amorous propensities, will be likely not to have children; and much, too, to their own surprise, as they conceive that they are specially qualified to bring about a different result. They little dream that *haste* is not *speed*, and that a slower pace would be a readier means of getting to the end of their journey. In other words, there is too much intensity in their devotions; for what they *produce* they *destroy*. Time, however, by qualifying their ardor, usually remedies this defect. Again, conjugal enjoyment should be followed by repose, on the part of the *female*, as but very little motion or agitation in females of warm temperaments is sufficient to arrest the *ovulum* on its way to the place assigned it. And furthermore, when it is supposed conception may have taken place, the parties would do well if, for at least a month, they would put a bridle on their desires; as the spasmodic embrace of a very amorous couple is calculated to disturb the embryo in its earlier state of existence, and hence, to occasion abortion or miscarriage."

WEAKNESS OR DEBILITATION.

" But, say Baillie, Swammerdam, Larry, and Dubois, the leading cause of sterility is weakness or debilitation, on the part of the male or female, or both ; and, adds the last, if this matter were duly attended to, nine-tenths of the people who are now pining for heirs *might be blessed with numerous progeny.*

" This weakness or debilitation may be in some instances natural; but it is generally an artificial result, produced by severe labor, libertinism, long residence in an unhealthy climate, secret habits acquired in youth, and other causes. This is the case when the man is in fault; and to the same

causes, or nearly such, may be attributed the deficiency in the other sex. Immoderate love of dancing and tight lacing may also produce an artificial
sterility in woman, by causing a looseness and lassi-
tude of system—the consequence of which is an
inability to respond to the action of the male, by the
sympathetic pressure which is necessary for the
conveyance of the *ovulum* to the chamber prepared
for its reception and nourishment.

" Debilitation, in man, is mostly owing to derangement of that part of the system connected with the
functions of generation. The result of this is, that
the semen is not of a teeming and busy nature, and
that it is not imparted with a force sufficient at
the site of fecundation. Hence, a stimulant is necessary; but it is also necessary that this stimulant
should strengthen without much excitement, or the
latter evil only will be corrected; that is, the requi-
site force will be obtained, but not the nourishment
required by the reproductive principle. Indeed, it
is known that drugs of merely exciting qualities
rather impoverish this principle than otherwise;
which accounts for the disrepute into which Spanish
flies, arrow-root, etc., have fallen in late years.
Water-cresses, duckweed, carrots, dandelion, arti-
chokes, figs, potatoes, shell-fish, peaches, eggs, and
oysters, are no doubt incentives to amorous propensities; but whether their influence extends farther, I
am in doubt. Again, the females of some countries
swallow flies, spiders, ants, crickets, and even frogs,

to promote fecundation; and in Spain they sip dew
from the olive leaves for the same purpose."

THE LONGING FOR OFFSPRING.

Dr. Foote says, in a chapter on barrenness:
"This word designates a physical condition abhor-
rent to every one not already a parent. Whether
love of children is limited or universal, it is one
from which every individual who has been long
married, and has not at least one child to enliven
the family, instinctively recoils. Such a condition
has in all nature but one parallel, and that the great
desert which spreads its vast expanse weariedly before the eye, without a blade of grass, leaf, twig, or
tree to nod a welcome to the passing breeze, nor the
first crystal of water to reflect its prismatic colors in
the golden rays of the sun. By many females the
grave is more cheerfully looked forward to than
childless longevity; and not a few husbands would
rather die in the prime of manhood, leaving an heir,
than to live to gray old age and be esteemed incapable of reproduction. The careless world can not
know the secret yearnings of the hearts of such unfortunate persons.

" A lady who has had four or five children generally wishes herself barren, feeling that she has done
her share toward populating the world; and she is
entirely unfitted by her fruitfulness to sympathize
with one who, loving children, has none of her own

to love. But, taking a serious view of the matter, however badly children may sometimes turn out, childless old age is a dismal future for the mind to dwell upon; and, having reached it, the present is no less cheerless. The hearthstone of a married pair, in the vigor of life, is electrified with the presence of the bright roguish eyes which mischievously watch the smiles and frowns of approving and reproving papas and mammas; while no vernacular is so enchanting as the hesitating and rambling utterances of 'our baby' when it first begins to kill the king's English. The new father seems more dignified, and stands several inches higher in his stockings, while the mother is never tired of relating the extraordinary feats and accomplishments, or quoting the wise remarks of her prodigy. Passing the meridian of life, doting parents watch with pride the developing genius of a promising son, or the unfolding brilliancy, beauty, or goodness of a favorite daughter; while the infirmities of old age are deprived of their depressing influences by the affectionate attention of grateful children.

"Therefore, the desire for children is natural, and all honorable means to obtain them excusable. A lady devotedly attached to them can not imagine how far she might go in her attempts to become a mother, unless placed right in the position of one who has spent many years of married life without a sign of pregnancy. Nor can a man picture to himself the mortification one feels who passes years in the state of matrimony without becoming a father, unless he has himself been in that position."

BARRENNESS NOT CONFINED TO EITHER SEX.

" Barrenness is not always peculiar to females, as many suppose. Wives are often considered barren by their husbands and friends, when all they require for fecundation is the introduction into the vagina or womb of a healthy spermatozoon; or, in other words, one of the animalculæ found to exist in the healthy male semen, possessing vitality enough to impregnate the ovum or egg of the female. A husband may be to all external appearances healthy, and his procreative organs may seem perfectly sound and capable of yielding sexual gratification to both himself and wife; but if the seminal fluid which he deposits in the vagina is destitute of spermatozoa, or if the latter are feeble and almost lifeless, then in reality he is barren. Feeble spermatozoa have not the requisite animation to ascend the cavity of the womb or Fallopian tubes; or, having done so, they have not the strength to penetrate the egg.

" It is undeniable that barrenness is more often the fault of the wife than of the husband. There is probably no *good* reason why this should be so, except the fact that attention to domestic duties renders the faithful housewife more sedentary in her habits."

CAUSES OF BARRENNESS.

"Not by any means one of the least of avoidable causes is the way in which the young girl grows up, without learning any thing of the care necessary to be bestowed upon her sexual organs, except what she stumbles into or learns accidentally. The dawn of puberty is an important epoch in the life of the young female; and at this time she needs advice which only the intelligent mother or the physician can give. All inattention to the natural functions, or a local derangement of the procreative organs, has a tendency to induce nervous and vascular derangements, and these in turn fasten the local difficulty more permanently upon the victim.

"The most common immediate cause of barrenness in the female is a diseased condition of the ovaries, where the eggs or ova are produced. There are a great many names given to the various forms of disease these organs are subject to, but they may be briefly summed up as inflammatory, ulcerous, cancerous, tumorous, dropsical, or paralytic in their character.

"Leucorrhœa and ulceration of the womb are often immediate causes of barrenness. If the secretions of the vagina, or the discharges from the ulceration of the womb, are of a very acrimonious character, the spermatozoa of the male, in their ascent toward the Fallopian tubes, might as well encounter the poisonous secretions of the Upas-tree as those

4

deadly fluids from organs which, in health, extend a hospitable welcome to them.

"Barrenness in females may be caused by the Fallopian tubes having become glued to the ovaries. It may be produced by suppression of the menses; by the induration or relaxation of the neck of the uterus; by the want of erectile or contractile power of the procreative organs of the female; by too great similarity in the temperaments of the husband and wife; by too great amative excitability, at time of coition, in consequence of which the violent movements of the Fallopian tubes rupture the egg or ovum; or an excess of reciprocal magnetism destroys the vitality of the male and female germs."

UNFRUITFULNESS IN MEN.

"Obstinate barrenness in males is more difficult to cure, and sometimes baffles the skill of the physician. Strange as it may appear, the artificial injection of healthy male semen into the vagina has been resorted to by resolute and determined, but virtuous, wives in their childless despair. Some physiologists claim that the spermatozoa of the male will retain their vigor and impregnating power if put in warm water, and injected with a syringe; but no successful experiment is adduced to sustain the hypothesis. Still there are means by which the artificial injection of healthy male spermatic fluids may be made so as to induce impregnation. Electricity,

however, properly applied, will cure many a case
of male barrenness, which ordinary systems of med-
ication have failed to favorably effect. No married
pair should despair of having children. Too much
care to protect the embryo can not be taken by a
lady who, after years of fruitless marriage, becomes
enciente. Such a person is much more liable to
miscarry; and miscarriages are apt to render a pre-
disposition to barrenness more confirmed."

CHAPTER IV.

THE MICROSCOPE IN THE TREATMENT OF STER-
ILITY.

DR. J. MARION SIMS, in one of the most practical and valuable contributions to science, read
before the Medical Society of the County of New York,
in December, 1868, elaborates the wonderful value
of the microscope in the diagnosis and treatment of
the sterile state. His paper is so full of important
suggestions, that we reproduce it here—abridging
and omitting some portions—for the benefit of our
hundred thousand readers who will never see the
original. He thus tells the secret of his great interest in, and thorough application of, the microscope:

" About ten years ago, my friend Dr. W. H.
Dwinelle, a very accomplished amateur microsco-
pist, was showing me some organic substances under
the microscope, in which I did not take any great
interest. Indeed, I felt positively bored by the per-
tinacity with which he attempted to enlist me in his
favorite study, when, at last, he said: 'My dear
doctor, I see you are very tired; but here is something you must look at before you leave.' He held

the glass up, to show me there was nothing my eye could detect, then placed it under the microscope, adjusted the focus, and asked me to look. I am sure I was never more surprised in all my life than I was then, to see and read the Lord's Prayer. From that moment I was convinced that the instrument exhibited things just as they were. And I have often said that I supposed this was not the first instance in which the Lord's Prayer was the means of opening a man's eyes to the truth."

INVESTIGATION OF STERILITY.

He then proceeds to say that " in the investigation of sterility there are three questions that must be settled at the outset, if we expect to treat it understandingly.

" 1. We must be sure that we have semen with spermatozoa.

" 2. We must ascertain if the spermatozoa enter the utero-cervical canal.

" 3. We must determine whether the secretions of this canal are favorable or not to the vitality of the spermatozoa.

" For, if the semen does not contain spermatozoa, of course the uterine condition does not call for any treatment whatever. But if it does contain spermatozoa, and if they do not enter the cervical canal, then there is the question of operation or not, to permit their entrance.

" On the other hand, if we should find spermatozoa in the cervical canal, then, as a rule, no operation will be needed; and if we should find them there in abundance, and all alive, then the case needs no treatment whatever. But, if we should find them there, all, or nearly all, dead, then it is evident that the secretions of the utero-cervical canal poison them, and therefore the physical condition, giving rise to this abnormal secretion, must be searched out and treated."

WHEN CONCEPTION IMPOSSIBLE.

" The present advanced state of physiological knowledge warrants us in saying that conception is impossible without spermatozoa; and that it is impossible if the spermatozoa can not pass into the cavity of the uterus; and, to these acknowledged truisms, I must add another, viz., that it is equally impossible if they die in the cervical canal, or are dead when they reach the uterine cavity. It is, then, self-evident that these three points must be determined—it matters not what other complications may exist. Fortunately, as I have said before, they are all easily and quickly settled by the microscope. Without the microscope, it is impossible to determine either of them. Without the microscope, then, our treatment of the sterile state is simply blind empiricism. With it, our diagnosis becomes absolutely certain, and our treatment, at least, rational.

What, then, are the first steps in this investigation? How can we begin? Where shall we begin? Now, as it is upon this very point that I have been so stigmatized, I will tell you exactly how I manage this delicate affair."

MODUS OPERANDI.

"Given a case of sterility for investigation, the physician examines the state of the uterus and its appendages. His patient may have a frightful dysmenorrhœa, a flexed cervix, a contracted cervical canal, some malposition, a polypus, a fibroid, or something that would possibly prevent the passage of the semen to the cavity of the uterus. He may feel convinced, in his own mind, that conception can not take place unless some surgical operation be performed—perhaps incision of the cervix uteri. This operation is usually done to permit the passage of the semen into the uterine cavity. But in this, or in any case, what right have we to say that the semen does not or can not pass into the cavity of the uterus? We must not take it for granted that it does not, simply because the os seems to be small; for we know that cases are recorded where conception occurred when the os barely admitted a small-sized probe; and we know very well that spermatozoa now and then pass along the Fallopian tubes, which ordinarily admit a bristle. If the semen enter the cervical canal, we may lay it down as a rule, that a dilatation of the cervix by incision, or otherwise

is not necessary ; but if it do not, it may be necessary. We perform any rational operation for the
relief of suffering, and for the restoration of health ;
but I insist that we have no right to perform any
operation, or to institute any treatment whatever,
solely with the view of the cure of the sterile condition, till we have first settled the three propositions,
above laid down, touching the presence and viability
of the spermatozoa. To find out all at once, and
with the least delay and trouble, I usually say to the
husband or wife, as it may be, ' It is very important,
before instituting any treatment, to be sure that
the seminal fluid enters the neck of the womb, for
without this conception is impossible. We must also
ascertain if the uterine secretions kill the semen ; if
so, a certain treatment will be necessary. If you
will, then, send your wife here, or come with her
any day, five or six hours after coition, it will be
easy to settle these points at once.' In nineteen
cases out of twenty, the wife presents herself the
next day. The speculum is introduced (and when I
say the speculum, I always mean the one that bears
my name), and some vaginal mucus is removed by
the syringe, and placed on an object-glass. Then
some cervical mucus is drawn out and placed on
another object-glass. These two specimens are then
examined under the microscope. If we find spermatozoa, well and good ; but if we find none, neither
in the vaginal, nor cervical mucus, our fears are at
once aroused. What, then, is to be done ? I simply

say that I am not quite satisfied with the examination, and would like to see the wife again, at some future time, under the same circumstances. But, suppose we find no spermatozoa on this second examination. Then two questions immediately arise: either, that there are no spermatozoa, or that the semen has all passed off before the case came under observation. Sometimes the semen is all instantly thrown off by the vagina, and then it would not do to pronounce the husband sterile till we are sure of a specimen of his semen, for investigation. If I fail to satisfy myself on this point, I then explain the possibility of the semen all passing off, in the act of rising and dressing, and show the absolute necessity of making the examination half an hour or so after coition, and before the erect posture is assumed. When the subject is presented in this plain, practical manner, and treated seriously, no man or woman of sense could oppose it; and with me, it has never, in a single instance, been objected to. When I am sent for to make the examination, if I find in the vagina a fluid with the characteristic seminal odor, I am satisfied with the microscopic examination. I have never, in but two instances, been compelled to resort to Mr. Curling's plan, of asking a man to squeeze a drop of mucus from the urethra, on to a bit of glass, immediately after sexual intercourse. But as this is sometimes necessary, it is well to remember it.

" If we eventually find that the semen contains no

spermatozoa, then all uterine treatment is at an end.
But if we are at last satisfied that it contains spermatozoa, then we must determine if these enter the
cervix uteri; and if so, do they there find a fluid
favorable to their existence alive? And all this can
be done only by the microscope.

"The question of the entrance of the semen into
the cervical canal, and of the effect of its secretion
upon the spermatozoa, can be fully and satisfactorily
ascertained only during a very brief period. We are
sure to make a mistake if the microscopic examination be made just before the expected return of the
menses; and why? Because there is always a certain amount of fullness of the uterus—of engorgement,
so to speak—which precedes the menstrual flow; and
the cervical canal may not admit the semen from
mere turgescence of its walls. Besides, at this time,
its secretions are almost sure to kill the spermatozoa
even if they should happen to enter this canal."

WHEN CONCEPTION TAKES PLACE.

"Physiologists are generally agreed that conception takes place during the week following menstruation. Avrard says we have fourteen days of
active uterine life and fourteen days of uterine hyp-
notism. He says that conception can occur at any
time after menstruation up to the fourteenth day,
counting from its commencement. For instance, if
menstruation should last for three days, then we

would have eleven days for the possibility of conception. But, if menstruation should last eight or nine
or ten days, then we would have respectively but six
or five or four days as the time possible for conception. After this time the uterus, according to Avrard, lapses into the hypnotic state, when conception
is impossible. While I am disposed to accept Avrard's dictum as the rule, I think I have seen exceptions to it, if we can always depend upon testimony
seemingly reliable. Be this as it may, I am sure of
this fact: if we wish to determine the effects of the
cervical mucus upon the spermatozoa, we must make
the experiment during the week that follows menstruation. About the fifth or sixth day after the
flow is the best moment; for then the uterus is in the
most favorable condition. The cervical mucus, which
just before menstruation was perhaps thick and
opaque, then becomes clear and translucent. If, by
examination made at this particular period, we should
find spermatozoa in the cervical mucus, we could
safely say that it will not be necessary to incise the
neck of the uterus. But if the sperm do not enter
the canal, then the probabilities are in favor of the
necessity of some surgical interference. The semen
may enter the cervix in great abundance, and we
may find the spermatozoa all dead, even but a few
minutes after coition. Then, as said before, we must
find out the source of the poisonous secretion and
remedy it; for a vitiated secretion shows some organic condition requiring a special treatment."

EXAMINATION OF SPERMATOZOA.

" When I wish to examine the action of the cervical mucus upon the spermatozoa, I order sexual intercourse in the morning—the dorsal decubitus to be retained for an hour afterward; and I expect a visit from my patient four or five or six hours after coition. Sometimes we find spermatozoa in great abundance in the cervical canal, and not one living. (I have occasionally examined the mucus six, eight, and ten minutes after coition, and found all the spermatozoa dead.) Sometimes we find half of them dead; again, only about a third; again, two-thirds. Sometimes, in one portion of the mucus, every spermatozoon is dead, while, in another portion of the same sample with fewer epithelial scales, we may find them alive.

" Now and then, after treatment for a month or more, I have found the mucus drawn from the lower segment of the cervical canal full of living spermatozoa, and I have supposed that the case was cured; but when I came to examine that drawn from the upper segment of the canal, near the os internum, they were nearly all dead. This was evidently because the mucous membrane lining the lower segment of the cervix, being more easily reached and more thoroughly treated, had assumed a healthier character, and consequently its secretion was restored to a normal condition; while that higher up, and more difficult to reach, had not been so much improved, and hence its secretion was still abnormal— a condition requiring further treatment.

"The vaginal mucus, by its natural acidity, kills very quickly every spermatozoon. I do not now remember ever to have found one alive in the vagina, except when the examination was made very soon after coition; when, indeed, the vagina was full of semen but slightly mixed with other secretions. Examined three or four hours after intercourse, the spermatozoa found in that portion of the mucus of the vagina adhering to its walls are always all dead. Indeed, the normal vaginal secretion seems to be a perfect poison for the superabundant spermatozoa.

"When, after a month's treatment, I wish to know whether the case is cured or not—in other words, whether all possible recognized barriers to conception are removed—I order sexual intercourse (just after menstruation) at night, and examine the cervical mucus twelve or fourteen hours afterward. If the majority of the spermatozoa be alive and active, I have great hope of conception. Before dismissing a case as cured, I think it necessary to examine the mucus thirty-six hours after coition; and, if it is then all right, of course I suspend the treatment, and patiently wait the hoped-for result."

MICROSCOPIC EXAMINATION.

"So much care is necessary in the removal of the mucus for microscopic examination, that I may be

pardoned for referring to it again. The patient is
placed in the left lateral semi-prone position, as I
have elsewhere so minutely described, and my speculum is introduced, and some of the vaginal mucus
drawn up with a small glass syringe, previously
washed out with warm water. This is deposited on
the object-glass; the vagina is then cleared of all
secretion, whether vaginal or cervical, the whole of
the vagina and the os uteri being thoroughly wiped
over with a pledget of cotton. This is for the purpose of guarding against the possibility of mixing
vaginal with cervical mucus, which would, of course,
spoil the whole experiment. The cervix is then
brought forward either by the depressor or a tenacu-
lum, if necessary, which enables us to look directly
into the cavity of the cervical canal. The syringe is
then to be again thoroughly rinsed in warm water;
its nozzle is passed into the gaping canal for half an
inch, and the cervical mucus in its lower segment is
drawn out. The instrument is emptied, washed out
again with warm water, and reintroduced up to
the os internum, and another portion of mucus is
drawn out, provided there is any left after the first
effort. Thus we have three specimens of mucus;
i. e., one vaginal, and two cervical. The cervical
secretion should be clear and translucent, and about
the consistence of the white of egg. If it contain
any little opaque specks of milky whiteness, it invariably poisons the spermatozoa to a greater or less
extent. We sometimes find the cervical mucus per-

fectly clear, and yet poisonous to the spermatozoa.
Here we would naturally expect to find excessive
alkalinity of the secretion; but I have not been able
to detect it. In these cases it has seemed to me that
the spermatozoa were killed—drowned, as it were—
by the very abundance of the secretion. I do not
here allude to cases of uterine catarrh, where the
secretion is very thick and albumino-purulent ; for,
of course, this is a deadly poison to the living principle of the semen. But I allude wholly to such
cases as have been changed by treatment to a condition giving rise to a secretion seemingly normal, so
far as an ordinary ocular examination is concerned.
Here the microscope is our unerring guide. The
mucus may be clear, and perfectly normal in appearance; but, if it kill the spermatozoa, then there is
still some point in the canal of the cervix, or in the
cavity of the uterus, that gives out a vitiated secretion ; and this must be found out and corrected before the case is wholly cured. When we find living,
active spermatozoa high up in the cervical canal
thirty-six or forty hours after coition, we can pronounce the case cured, so far as it can be by surgical
means, and not till then."

INCISION OF THE CERVIX UTERI.

" It is time for us to pause and consider if there is
not something more to be done for the sterile condition, than to split up the cervix uteri. I look upon

this operation as one of great importance, as one of
the most valuable in uterine surgery, but I think
that we have followed too blindly the example and
teachings of its illustrious author, Sir James Y.
Simpson. For myself, I am now sure that I have
cut open the cervix uteri, perhaps scores of times,
when it was both useless and unnecessary; and I
know that others have done the same thing. Do not
misunderstand me. I speak here solely of the operation with reference to the sterile condition, when it
would be wholly useless if the husband happened to
be sterile, and certainly unjustifiable unless imperatively called for by considerations of health. Incision of the cervix for dysmenorrhœa is one thing;
incision of the cervix for sterility, even if there be ,
dysmenorrhœa, is another, and it behooves us to
draw the line of distinction in every case, and not to
take it for granted that every woman is sterile who
may have dysmenorrhœa or feeble health, or that
every man is prolific who may be vigorous and enjoy
good health. I am sorry to say that I have had the
misfortune to incise the cervix in half a dozen cases
of sterility, where I found afterward, to my great
mortification, that the husbands were incapable of
procreation, because their semen had no spermatozoa,
and that, too, since I have known the value of the
microscope. In each case the operation was called
for to restore health, but was totally useless for the
relief of its incidental accompaniment, sterility, and
would not probably have been submitted to for con-

siderations of health alone, had it not been for the
hope of offspring afterward. I made the mistake of
operating on these cases, because the social position,
moral character, and appearance of health in the husband, conjoined with the excessive dysmenorrhœa
and utter prostration of the wife, led me to operate
without the preliminary step of ascertaining whether
there were spermatozoa or not. I wish others to
profit by my mistakes; and I am less ashamed to
tell you of them, than I am to own them to myself.
However, this can never happen to me again, and
should not, after this warning, happen to any of my
brethren. I know many men who have no spermatozoa, and can not, therefore, become fathers. They
are all strong, active men, in the prime of life, and
all perform the sexual function with vigor. The
very fact of their natural vigor and strong passions
had been their ruin, for most of them had contracted urethritis during their early and unmarried
life, and had suffered from its unlucky sequence,
epididymitis. To further illustrate the necessity of
the microscope in this department of surgery, I shall
append a few cases drawn up as succinctly as possible : "

NECESSITY OF THE MICROSCOPE ILLUSTRATED.

"No. 1 had consulted two of the most eminent physicians in
England, and remained under the care of one of them for many
weeks. She said that during that time the neck of the uterus
was repeatedly cauterized. She got impatient, and went to
another physician, who told her that the caustic treatment she

had submitted to was worse than useless; and that a surgical operation was the only thing to be done. She consented to it, and he incised the cervix bilaterally. She did not conceive, and two years afterward went to Paris to see me. I found the uterus normal in all its relations, the os tinctæ and cervical canal sufficiently patulous. I explained to both husband and wife the importance of examining the cervical mucus four or five hours after coition. They returned the next day; the cervical mucus contained spermatozoa; therefore there was no necessity for any further surgical operation. But the spermatozoa were all dead; therefore there was a necessity for a treatment to rectify the vitiated cervical secretion. She remained in Paris a few weeks under my care, was cured, and became a mother in a year after her dismissal. Now, if the first physician had used the microscope, as I direct, he would probably have found that the semen never entered the cervix at all; and, if the second one had done the same thing, he would certainly have found that the mucus of the cervix poisoned the spermatozoa.

"No. 2, a lady, in the highest ranks of life, was sterile. The cervix uteri was incised bilaterally. She had pelvic cellulitis afterward. Two years after this I saw her, and she was still childless. The microscope showed that the cervical mucus, examined four hours after coition, killed all the spermatozoa. While this condition exists conception is impossible.

"No. 3, sterile, was treated for sterility in America, for a long time (two or three years). She went to Europe; had the cervix cut open, and was sent away with the promise of offspring. I saw her some time afterward. The microscope proved that the husband was sterile. Therefore, the previous treatment at home and the operation abroad were useless. I could relate several other cases like the above. But, as I have often made the same mistake before I fully understood the value of the microscope, I forbear.

"No. 4, married four years; sterile. She had dreadful dysmenorrhœa, followed by a discharge of a bloody brownish mucus, of an offensive odor. The uterus was anteflexed; anterior wall hypertrophied; os uteri small. I was in doubt, at first, whether to recommend an incision of the cervix or not. I told the husband that an operation would be necessary if the semen did not enter the canal of the cervix; but, if it did enter, the case might be cured without cutting. The wife came to see me the next day, some five or six hours after sexual intercourse. A drop of mucus from the cervix contained spermatozoa in great abundance. Here, the whole question of diagnosis and treatment was settled at once, and in the only way possible, by the microscope. For this one examination proved all that was essential to know—viz., 1. That the semen was perfect; 2. That it entered the cervical canal, and therefore there was no surgical operation necessary; 3. That the cervical mucus poisoned the spermatozoa, and hence a treatment directed to the utero-cervical canal was indicated. After the next menstruation (a month's treatment), the cervical mucus was considerably improved, for it contained large numbers of active spermatozoa. At the end of two months, I found living spermatozoa in the cervical mucus, thirty-six hours after coition. All treatment was now suspended, and after the next menstruation conception took place.

"No. 5, married five or six years, without offspring. The uterus was small, and retroverted by a fibroid, about the size of a walnut, on its anterior surface, just at the junction of the cervix and body. The os was very small, so small that a most distinguished accoucheur advised incision of the cervix, to admit the passage of the semen, although he was not in the habit of performing the operation, and, as a general rule, was opposed to it. In former years, I would have given the same advice without the slightest hesitation. But now I said, No. Let us first see if the cervix admits the semen. If so, the operation is hardly necessary. If not, it is. I saw the wife the next day. A drop of cervical mucus, under the microscope, determined the question against the operation at once; for the mucus was full of spermatozoa, but they were all dead. During the treatment of this case, I have seen the mucus in the lower segment of the cervix full of living spermatozoa, while that taken from the os internum was full of dead and dying ones. Nothing but the microscope could have revealed the truth in such a case as the above.

"No. 6, married eight years, sterile, had been treated by several distinguished physicians for the sterile state; and had been to Ems and other watering-places, all for no result. At last she came to Paris, to see my friend Sir Joseph Olliffe, and he called me in consultation. I found a long, conical, indurated cervix, with a small os—just such a case as I would have pronounced sterile by necessity, and just such as I have over and over again operated upon without further thought. But now I wished to be sure, before recommending an operation. After explaining the necessity for it, I requested this lady to come and see me, four or five hours after coition. She returned the next day. I could find no spermatozoa in either vaginal or cervical mucus. I requested her to come again. I saw her two days afterward—no sign of spermatozoa. I then told her that perhaps the seminal fluid all passed away in the act of rising and dressing. She thought it did. After further explanations, she readily agreed to send for me some morning, to verify the state of affairs. She was a very sensible woman, and fully understood the reasons given. A day or two afterward, I saw her in bed, about thirty minutes after sexual intercourse. The vagina was full of semen; and I removed about a drachm of it, and went home immediately for the microscopic examination. But, unfortunately, there were no spermatozoa. Not very long ago (seven or eight years), I had the idea that sterility was essentially a female infirmity; that men were never sterile, except when impotent; and that any man, legally competent for the married state, was physically so for procreation. But the microscope unsettles and settles all such vague notions. It is natural to suppose that a strong, vigorous man is more fitted for procreation than a weak or puny-looking one. Some of the greatest lights of the profession have held such views as this. It was only two or three years before the death of the lamented Trousseau, that he said to me, in speaking of a case we had under consultation, 'If our patient only had a man for a husband, all would be right.' I subsequently found out that the husband's passions were strong; that his semen was perfect; that it entered the cervix in great abundance; and that the spermatozoa were there poisoned by a vitiated secretion. I mention this to show that we must not judge from appearances, when it is so easy to settle the question by the microscope.

"No. 7, married nine years, sterile, had consulted several distinguished physicians, one in Germany, who told her that it was useless to try any further treatment, as she was now well enough, and that it was the fault of her husband that she did not conceive. I explained to her that there was nothing easier than to determine that question at a single visit. She came the next day. I removed some vaginal mucus; also a mass of cervical, as large as a pea, that was just hanging from the os; also some from within the canal. The vaginal mucus contained spermatozoa, but, of course, they were all dead. The mass of cervical mucus that hung out of the os contained spermatozoa in abundance, all dead. The mucus from the interior of the cervix was wholly devoid of spermatozoa."

"Here the microscope settled the whole question. There was no longer any guess-work. 1. It was not the fault of the husband that there had been no conception. 2. The seminal fluid did not enter the canal of the cervix. 3. The spermatozoa were killed by the cervical mucus, where the two came in contact. As the shortest and best method of treatment, I incised the cervix. After the subsequent menstruation, semen was found to enter the canal of the cervix. After the next period, they were found there in abundance, and all living. In three months thereafter, she conceived. In another three months, she miscarried, in consequence of a fall. Six months after this, she conceived again; and a year ago she became a mother.

"So far I have related only cases of natural sterility,

and, were it necessary, I could give scores more of the like character, but, there is so much sameness among them, that it would be superfluous. However, bear with me, while I give one or two illustrations of the value of the microscope in acquired sterility."

THE MICROSCOPE IN ACQUIRED STERILITY.

"No. 8, aged 36, had given birth to one child, ten years ago. Her general health was perfect, but she did not conceive again. She was anxious for more offspring—had been to various watering-places, and had consulted several distinguished physicians. At last she fell into the hands of my friend Dr. Lheritier, who brought her to me. I found the uterus hypertrophied, and somewhat retroverted. The os was rather small and the cervix indurated, and I had some doubt whether the semen could enter the cervical canal. But a microscopic examination proved that it did, and that the cervical secretions killed all the spermatozoa. This case was under treatment in January and February, and again in May and June. When she left in June, living spermatozoa were found in the cervical mucus, in great abundance, thirty-six hours after coition. We, therefore, pronounced the case cured. She conceived a month afterward, and was safely delivered at term.

"No. 9.—We often fail to cure curable cases because the treatment is sometimes so tedious that both patient and doctor get mutually tired, and both are glad to quit. Madame ——, aged 34, had one child eight years ago; subsequently had chronic cervical inflammation; was cauterized too much. The cervix became indurated, and the os contracted. She wanted more offspring. I was in doubt about cutting open the cervix. A microscopic examination proved that the semen could not enter the cervix. Accordingly, I incised the os. After this, the semen entered the canal of the cervix, but its mucus killed all the spermatozoa. The mucus was not as clear and limpid as it should be, and it had white milky specks in it, looking as if it had been mixed with a little of the vaginal secretion. The lining membrane of the cervix was too red and rather granular. This was cauterized even up to the cavity of the uterus; and various other local as well as general remedies were adopted and carried out from time to time for twelve months. The character of the cervical secretion gradually improved, and at times showed some living spermatozoa, and again all were dead. This patient did not despair, notwithstanding a fruitless treatment for so long a time.

"A sponge-tent had revealed long ago a small flattened cystic tumor in the canal of the cervix, situated on its posterior face, just at the os internum. I had repeatedly suggested the propriety of extirpating it. After all other means had been exhausted for restoring the cervical secretion to a normal state, the operation was agreed to. In June, 1867, nearly two years after we began the treatment, a sponge-tent was introduced; the canal of the cervix was fully dilated, and a cystic tumor, about the size of the end of the little finger, was extirpated. Three months afterward, the cervical mucus was greatly improved; and in March last, after a treatment of more than two years and a half, I examined the secretions fifteen hours after sexual intercourse, and I had the satisfaction of saying, 'At last, madam, I find the cervical mucus perfect; it is full of spermatozoa, and all very active. We can now hope for conception.' Conception dated from that period, for she did not menstruate afterward. But for the microscope, I would have dismissed the case as cured after the incision of the cervix uteri, and she would have remained, in all probability, sterile to the end."

" Once I thought that the most common obstacle to conception was a contracted cervical canal, contracted at its outlet, at the os internum, or throughout its entire length. But, if I were now asked, 'What is the most frequent obstacle to conception?' I should unhesitatingly say, 'An abnormal utero-

cervical secretion that poisons or kills the spermatozoa.' I can call to mind numbers of cases where,
in former years, I incised the cervix, when the operation was satisfactorily done, and yet the sterility persisted. In some of these I have now not the least
doubt that the husbands were sterile, and in others
I have as little doubt that the cervical mucus was
poisonous to the spermatozoa. If I had then possessed the exact knowledge of to-day, how much
more satisfactory would it have been for me—how
much better for my poor patients!

"I could go on for hours with cases to illustrate
the principles already laid down. The foregoing are
taken at random, and are sufficient for the purpose.
I have not treated a single case of sterility as such,
in the last six years, without determining the three
questions so essential to success that were stated at
the outset of this paper, except the half-dozen cases
already alluded to ; and in these the miscroscope at
last revealed the truth."

EXTENT AND CAUSES OF STERILITY.

The same writer, Dr. J. Marion Sims (in *Am. Jour.
Med. Sciences*, Oct., 1865, p. 557) said, at the then
recent meeting of the British Medical Association,
that every eighth marriage was sterile. Many of these
were caused by misplacements of the uterus. Of 250
cases of " natural sterility " (*i. e.*, of those who were
married a sufficient length of time, and did not con-

ceive), 103 had anteversion, and 68 retroversion; and of 255 cases, that had fallen under his observation, of acquired sterility (*i. e.*, where women had borne children, but had ceased to do so long before the termination of the child-bearing period), 61 had anteversion, and 111 retroversion. A number of these cases he had cured—so that the patients, after six, ten, and even fifteen years of sterile marriage, had borne children.

SINGULAR CURE OF MALE STERILITY.

The French *Journal de Med. et Chir. Prat.*, 1866, records, from the pen of M. A. Amussat, jr., a case of sterility cured by removing a phimosis. The fact of five years marriage without his wife becoming pregnant, led to the discovery, on examination, that the gentleman had a very contracted phimosis with excessive length of prepuce, so that the gland could not be uncovered; and when he urinated, the præputial sac became filled like a funnel, from which the urine afterward flowed in a very thin stream. May 11, 1866, M. A. removed the prepuce by circular cauterization (circumcision with the knife would have been better). The part separated on the 25th, and in July the cicatrization was complete, and the remaining prepuce could be drawn back so as to uncover the gland, and urination became free. One year afterward the gentleman's wife gave birth to a son.

CHAPTER V.

AGE AT WHICH MENSTRUATION IS ESTABLISHED.

IT seems well understood that the most singular, curious, and interesting sensation experienced by the female sex, prior to the pangs of childbirth, and the joys of motherhood, is that of menstruation—that internal purification and radical change of system which makes her, for the first time, capable of child-bearing. A number of curious phenomena accompany this development: besides the change of voice, the throat and neck become larger and more symmetrical; the breasts swell, and nipples protrude; the chest expands, and the shoulders become rounder and plumper; the eyes brighten, and fairly sparkle with intelligence; in fact, the girl becomes a woman, almost a new being. The age when this occurs varies from ten years, in warm climates or near the equator, to eighteen and twenty years in northern latitudes. In the United States, fifteen is the age at which much the larger number of girls arrive at puberty; nearly as many *before* as after that age.

The following table contains only well-authenticated cases, together with the authority for all or most of them:

Number.	Age at which Menstruation was established.	Age of Conception, but where nothing is said of Menstruation.	Authority for the Case, and other particulars.
1	In 9th yr.		Montgomery on Preg., p. 256. Menses regular, with mammæ, and other evidences of puberty fully developed.
2	Just 12 years.		Montgomery on Preg., p. 88. A little girl; menses delayed by imperforate hymen; when punctured, discharge large.
3	10 to 12 years (many).		Montgomery on Preg., p. 256, says he himself had known many cases of menstruation between those ages.
4	4½ years.		This extraordinary case is mentioned by Sir Astley Cooper, Med. Chir. Trans., vol. iv, p. 490. For reference to several other cases, see Davis' Obst. Med., p. 236, and Beck's Med. Jour., p. 368.
5–14 15–33 34–88	10 in 11th year. 19 in 12th year. 55 in 13th year.		Part of a registry of 450 cases, kept by Mr. Roberton, of Manchester, England. (Edinburgh Med. and Surg. Journal, vol. xxxviii, p. 231.) Of these, three, grandmother, mother, and daughter, had become regular at twelve; and five sisters in one family menstruated at eleven.
89		10 yrs. 6 mos.	This is remarkable as the earliest instance of pregnancy, satisfactorily authenticated, that ever took place in Great Britain, the girl being only a few months over eleven at time of delivery. She had menstruated before she became pregnant. (Roberton on Midwifery, p. 30.)
90	10 yrs. 6 wks.	11 yrs. 9 mos.	This girl, named Sprayson, preferred a charge of rape against her uncle, who was convicted of the crime, four times repeated, at intervals of a week, when she was only eleven years and nine or ten months old. A child, full-

Number.	Age at which Menstruation was established.	Age of Conception, but where nothing is said of Menstruation.	Authority for the Case, and other particulars.
91	1 year.	9 years 3 mos.	grown and healthy, and after short and favorable labor, was born when she was twelve years and seven months old. (*British Record of Obs. Med.*, etc., Nov., 1848, p. 359.) This is the most remarkable case on record. Sally Deweese, born April 7, 1824, in Butler County, Kentucky, began to menstruate at a year old, and the pelvis and breasts became developed in an extraordinary degree. She continued to menstruate regularly up to 1833, when she became pregnant; and, on April 20, 1834, when only ten years and thirteen days old, she was delivered of a healthy female child, weighing seven pounds and three-quarters. (*Transylvania Med. Jour.*, vol. vii, p. 447.)
92		14 years.	Before she was fifteen years old, this girl had twins. (*Montgomery on Preg.*, p. 259.)
93		12 years.	This girl was delivered before she was thirteen, and had never menstruated. (*La Motte, Traité des Accouchemens*, p. 52.)
94		12 years.	Both these were known to Sir E. Howe, and reported in *Philos. Trans.* for 1819, p. 61.
95		11 years.	
96		9 years.	This case was known to Dr. Goodeve, at Calcutta, and the other heard of by him, the child born alive in each case. Both were Hindoo girls. Dunlap frequently saw in Bengal, and Bruce in Abyssinia, mothers of eleven years of age. (*Beck's Med. Jurisp.*, p. 135.)
97		8 years.	

No.	Age	
98	Just 12 years.	This was an English lady, married on her thirteenth birth-day, and had seven children before she was twenty-two years of age. (*Montgomery on Preg.*, p. 258.)
99	10 to 12½ years.	During the year 1816, several girls, who had conceived at these ages, were admitted to the *Maternité* hospital at Paris, France. (*Med. Jurisp.*, vol. i, p. 257.)
100	8 years.	A case of parturition at nine years, in England. (German *Ephemerides*, dec. 3, an. 2, p. 262.)
101	12 years 9 mos.	A mulatto girl at Philadelphia, delivered at thirteen years and six months, who had never menstruated. Recorded by Dr. W. T. Taylor.
102	7 years.	Related by Madam Boivin. The subject commenced to menstruate at seven years, and did so regularly after her tenth year.
103–163	60 in 14th yr.	
164–236	72 in 15th yr.	
237–291	54 in 16th yr.	Part of a registry of 326 females, the ages at which they began to menstruate, published in the *North of England Medical and Surgical Journal.*
292–340	48 in 17th yr.	
341–360	19 in 18th yr.	
361–379	18 in 19th yr.	
380–384	4 in 20th yr.	
385–6	2 in 13th yr.	
387–9	3 in 14th yr.	
390–413	23 in 15th yr.	At the Lying-in Hospital of Christiana, Norway, for 1819, a record was kept of these 122 females. The mean age was 17.7—which is almost exactly the same mean obtained, at the same hospital, the year before. Of these, 3 menstruated every fortnight, 27 every three weeks, 83 every month, 3 every 5 weeks, and 6 irregularly from 4 to 8 weeks. The duration of discharge was: in 45 from 2 to
414–445	31 in 16th yr.	
446–466	20 in 17th yr.	
467–481	14 in 18th yr.	
482–493	11 in 19th yr.	

Number.	Age at which Menstruation was established.	Age of Conception, but where nothing is said of Menstruation.	Authority for the Case, and other particulars.
494-502 503-506 507	8 in 20th yr. 3 in 21st yr. 1 in 24th yr.		3 days, in 24 from 3 to 4 days, in 18 from 5 to 8 days. The mean (of the 87 thus reported on) was 3½ days.
508	1 in 3d year.		In March, 1868, there was exhibited, publicly, in Cincinnati, Ohio, a "baby woman," named Sophia Gantz, who, at four months over two years of age, had the full and fully-marked breasts, form, and menses peculiar to a grown woman. Dr. H., commenting upon this remarkable case, says that besides the cases mentioned by Tulpius, Saxonia, and other old authorities, Lieberg gives the case of one only one year old; Deckers, one of two years of age; A. Cooper, of one between two and three, and another of about the same age; Velpeau, also, gives the particulars of a young girl at Havana, in which this physiological function was established at the age of eighteen months, and continued afterward with normal regularity; and likewise refers to a similar case found among the recorded cases of Meckel, for 1827, and republished in an English journal, in which the establishment of this function took place at the age of only nine months, with corresponding physical developments. Dr. Francke records cases where the courses appeared in children as early as four years of age.
509-512	4 in 4th yr.		

Dr. Copeland reports a table of 1,604 females, of whom 10 menstruated at 10 years of age; 47 at 11 years of age; and 174 at 13 years of age. The average range he regards as between twelve and nineteen years.

In the Western General Dispensary, London, where 10,000 cases annually are treated by its officers, Dr. Bennet made, between July, 1846, and March, 1849, a record or synopsis of 255 cases of menstruation, with the result as stated in the margin.

Dr. J. G. Wilson, of Glasgow, records the case of a girl who, at the age of 13 years and 6 months, had a full-grown female child. She must have conceived at 12 years and 9 months.

Dr. Roberton (in his *Midwifery*) records the case of a factory girl who became pregnant in her 11th year.

Dr. Carson, of Montgomery County, Pennsylvania, reports two cases of delivery of child at 14 years 8 months; also several cases under 16; in all such cases conception took place before marriage.

In January, 1860, a negro girl, just 11 years, 8 months, and 18 days old, at Pulaski, Tennessee, had a female baby— both living and in perfect health.

513–222	10 at 10 yrs.
523–579	47 at 11 yrs.
580–753	174 at 13 yrs.
754–758	5 at 10 yrs.
759–773	15 at 11 yrs.
774–801	28 at 12 yrs.
802–836	35 at 13 yrs.
837–877	41 at 14 yrs.
878–921	44 at 15 yrs.
922–948	27 at 16 yrs.
949–969	21 at 17 yrs.
970–992	23 at 18 yrs.
993–1002	10 at 19 yrs.
1003–1008	6 at 20 yrs.
1009	1 at 12 yrs.
1010	1 at 10 yrs.
1011–1012	2 at 13 yrs.
1013	1 at 10 yrs.

Number.	Age at which Menstruation was established.	Age of Conception, but where nothing is said of Menstruation.	Authority for the Case, and other particulars
1014	1 at 9 yrs.		The Cincinnati *Lancet*, March, 1863, records the delivery of a girl, 10 years, 8 months, and 17 days, of a living child.
1015	1 at 10 yrs.		In Howard, Wright County, Minnesota, on December 28, 1869, a little girl, named Penolia R. Wilkins, just 11 years old, gave birth to a child weighing 7 pounds 5 ounces. Several months after, when the babe had increased to 15 pounds in weight, the child-mother weighed only 80 pounds, and was only 4 feet 8¾ inches high.
1016	1 at 10 yrs.		A little girl in the State of Tennessee, at the age of 11, gave birth to a child by her own grandfather, who was convicted and imprisoned for the crime.
1017	1 at 10 yrs.		Elizabeth Drayton was born at Taunton, Massachusetts, May 24, 1847. On February 1, 1858, when 10 years, 8 months, and 7 days old, she was delivered of a healthy male child, weighing 8 pounds, at full time. Circumstances go to prove that she was pregnant 24 days before she was 10 years old. Nine months after, both were hearty.
1018	1 at 10 yrs.		Dr. Tanner, in his work on Pregnancy, London, 1860, mentions the delivery, when the girl-mother was only a few months over 11 years, of a full-grown, still-born child. The girl's figure was that of a full-grown young woman; mammæ were fully developed, and she had menstruated before she became pregnant.

MENSTRUATION PROLONGED TO LATE PERIODS IN
LIFE.

It is interesting and sometimes important to know
the age at which menstruation ceases. The "change
of life," as it is usually called—or the time when it
ceases to be "after the manner of women," as the
Bible expresses it—is always a matter of deep con-
cern to a woman. Fortunate are those, as a general
thing, who pass this time while nursing their youngest child, and thus escape the pangs, and floodings,
and other ill-health which most of women suffer.
At or about the age of forty-five the great majority
of females experience this change; more of them,
probably, *before* than *after* that age.

The following table of ninety-five cases is made up
from authentic sources. It will be observed that the
cases numbered from 31 to 83, below, were those of
women who bore children at the ages named; it is
probable that some of them had a return of their
menses subsequently, while others may have again
become mothers.

No. of case.	Age of continued menses.	Authority for the Cases, and other particulars.
1	61	Menses still regular at her death, at 61. She had 32 children before she was 45, when her husband died. La Motte, *Traité des Accouchemens*, ch. xii, p. 71.

6

No. of case.	Age of continued menses.	Authority for the Cases and other particulars.
2	. 70	These 17 cases were noted by Mr. Roberton and Mr. Harrison, at Manchester, England. See *Edinburgh Med. and Surg. Journal*, vol. xxxviii, p. 254.
3–4	2 at 60	
5–18	14 beyond 50	
19	75	In this case, mentioned by *Gardien*, tom. i, p. 366, menstruation continued regular and healthy.
20–25	55 or 56	Cases at this age were known to Dr. Montgomery. See his work on *Preg.*, p. 259.
26	over 53	Bartholomew Mosse, in a petition to the House of Commons, in 1755, states that 84 of the women delivered under his care were between 41 and 54 years old; of these 4 were in their 51st year, and one in her 54th.
27–30	over 50	
31–43	12 in their 46th yr.	These 51 cases of childbearing occurred at Manchester (Eng.) Lying-in Hospital, and were part of 10,000 cases registered there; 385 of them were women between 40 and 45 years of age. As far as was ascertained the catamenia continued up to the period of conception.
44–57	13 " 47th "	
58–65	8 " 48th "	
66–71	6 " 49th "	
72–80	9 " 50th "	
81	1 in her 52d "	
82	1 " 53d "	
83	1 " 54th "	
84	54	At this age, in May, 1816, Mrs. Ashley, of Firsby, near Spilsby, was safely delivered of female twins. *Edinburgh Annual Register*, vol. ix, p. 508.
85	58	In France it was decided in Court that the rightful heir of an estate was a child proven to have been born when his mother was 58 years old. *Mem. de l'Académie de Chirurgie*, tom. vii, p. 27.
86–7	52 and 54	Knebel gives these two cases.
88–9	49 " 52	Two cases where the women were unmarried until 48 and 51 years old, when each became pregnant and

No. of case.	Age of continued menses.	Authority for the Cases, and other particulars.
90	51	had favorable labor. La Motte, *Obs.*, xcvi, and xcvii, pp. 189, 190. Dr. Labatt attended a lady married at 40, but who conceived for the first time, when past 50, and bore a living child. *Montgomery on Preg.*, p. 260.
91–2	60 and 63	Several cases are recorded, by respectable authority, of births at 60 and one at 63 years of age, but their accuracy is seriously doubted. In the English Court of Chancery the heirship to an immense property turned on the question whether a woman might have a child at 60 years of age, and it failed to be proved. *London Med. and Surg. Jour*, vol. iii, p. 686.
93–4	2 at 44	One lady was married for 19 years without any prospect of offspring, when at the age of 44 she had a perfect child, but never had a single drop of milk in her breasts; she never menstruated nor conceived afterward. The other lady was married 24 years before becoming pregnant, and bore her first and only child at 44.
95	94	The *American Journal of the Medical Sciences*, for Feb. 1831, mentions this case as recorded in the *Ann. Univ. de Med.*—that of a female in perfect health at the age of 94, who continued to menstruate from the 53d to the 94th year; her relatives were remarkable for their longevity.
96	47	A lady in Massachusetts, who had borne 10 daughters in succession, on May 26, 1870, gave birth to a son. She was 48 years old in Aug., 1870.

WHEN CONCEPTION TAKES PLACE.

It is a commonly received notion, among all classes, that conception takes place sometimes within a day or two preceding the menstrual flow, but, usually within a week or ten days after it ceases. While this is true as a *rule* (so true that many couples have for years prevented having children, simply by avoiding connection during that time), yet the *exceptions* are tolerably numerous. The following table of well authenticated cases is compiled from several sources, mainly from the great work of Montgomery, on Pregnancy. Of these, twenty-four conceived within ten days, while nine were delayed from eleven to twenty-five days. These nine were exceptional cases, and doubtless very rare.

No. of Case.	Woman's Age.	Date of Last Menses.	Date of Intercourse.	Intervening Period.	
1		October 29	Oct. 29		
2		April 10	April 10		
3	29	Sept. 23	Oct. 5	12	days
4	22	Nov. 12	Nov. 19	7	"
5	20	Sept. 26	Oct. 1	5	"
6		August 9	August 11	2	"
7		January 9	Jan. 10	1	"
8		July 17	July 22	5	"
9		January 10	Feb. 2	23	"
10		May 14	May 14		
11		Nov. 7	Nov. 12	5	"
12		Dec. 13	Dec. 13		
13		Nov. 6	Nov. 18	12	"
14		Nov. 7	Nov. 18	11	"
15		March 14	March 18	4	"
16		June 15	July 1	16	"

No. of Case.	Woman's Age.	Date of Last Menses.	Date of Intercourse.	Intervening Period.
17		August 15	Aug. 18	3 days
18		August 4	Aug. 6	2 "
19		March 5	March 12	7 "
20		Sept. 10	Sept. 15 16 17	5 to 7 "
21	26	Oct. 8	Oct. 18	10 "
22	27	Nov. 23	Dec. 3	10 "
23	23	May 15	May 22	7 "
24	28	Oct. 1	Oct. 10	9 "
25	25	Sept. 11	Sept. 18	7 "
26	29	Oct. 21	Nov. 1 to 7	11 to 17 "
27	31	May 24	June 18	25 "
28	29	Nov. 10	Nov. 11 to 17	1 to 7 "
29	30	Dec. 15	Dec. 25	10 "
30	33	Nov. 27	Nov. 28 29	1 to 2 "
31	25	Oct. 18	Nov. 10	23 "
32	26	Feb. 20	Feb. 22	2 "
33		June 28	July 19	21 "

STATE OF THE FEMALE SYSTEM DURING PREGNANCY.

Immediately on conception, the womb and all its appendages, or systematic parts, receive a remarkable increase of vital action. Innumerable little vessels connected therewith, which before conception were too minute even to convey blood, now become distended and begin to circulate blood freely. The tissue of the womb becomes infiltrated with serum (the thin transparent part of the blood), and its bulk is thereby increased, its texture softened, and its fibres separated—while, upon its internal surface, lymph is poured out, and the mucous membrane which lines the cavity is transformed. Subsequently, serum in large quantities is secreted, to form the liquor amnii, and the nerves of the womb so increase in number and size as to impart to it a more exalted degree of sensibility, which is quickly diffused through the system at large. "The virtue which proceeds from the male *in coitu* has such prodigious power of fecundation, that the whole woman, both in mind and body, undergoes a change." There is felt a sensation of feverish uneasiness, chills alternating with flushes of heat, sick stomach, disturbed sleep, languor, and sometimes drowsiness; menstruation is suppressed, and the breasts soon begin to evince an active sympathy, becoming swollen and sensitive; the pulse is generally quickened, the blood appears somewhat inflamed; venesection (bleeding) is resorted to in many cases for relief.

In consequence of these extraordinary changes and this increased vital action, the womb acquires a principle of growth which in a few months develops it from a small organ, almost hid among the contents of the pelvis, to a capacity more than five hundred and nineteen times greater. This growth, in its varying phenomena, has been not inappropriately styled the *miracle of nature.* The usual size of the womb of a virgin is, $2\frac{1}{2}$ to $2\frac{3}{4}$ inches long, $1\frac{3}{4}$ broad, and 1 inch from back to front; while the cavity is scarcely so large as the inner shell or wall of an almond; its superficies, or total outer surface, is estimated by Levret, at 16 inches. Just before childbirth, its length is 12 to 14 inches, its breadth 9 to 10, and from back to front it is 8 to 9 inches; instead of 16 inches, its superficies is now about 339 inches; and the tiny cavity of ⅔ths of a cubic inch will now contain about 408 cubic inches. All of its constituents grow similarly; for instance, blood-vessels which before conception would not have admitted the point of a probe will now readily receive the end of one's little finger. In a few weeks after childbirth, it will shrink to its original unpretending size.

This expansion of the womb necessarily affects other parts. As it grows, it gradually deserts the pelvis, and rises out of its cavity into that of the abdomen, thus disturbing the relations of all the contiguous parts. Generally, the first organ affected is the bladder, which becomes irritable, and occasions frequent urination. Toward the close of pregnancy, the female is often unable to retain her water except for a short time, and suffers much inconvenience by its coming away involuntarily while she walks, or if she coughs, laughs, or sneezes. Considerable uneasiness is felt in some of the connections of the organ, which extends also along the nerves of the thigh, producing numbness, cramp, and more or less pain along the limb; frequently, the power of one or both of the lower limbs becomes somewhat impaired—occasionally preceded or followed by dropsical swelling of the feet and legs. When the womb has acquired its full growth, it occupies a very large space in the abdominal cavity, pressing both the liver and stomach upward against the diaphragm—by which the capacity of the chest is diminished, the action of the lungs impeded, and much difficulty may be induced, sometimes attended with troublesome indigestion.

BODY-WEIGHT OF PREGNANT WOMEN.

Dr. Gassner, while at the Lying-in Hospital of Munich, Bavaria, in 1862, and previously, instituted an extended series of observations on the variations in body-weight of pregnant and lying-in women.

During the last three months of pregnancy, the body increases in substance a thirteenth.

The loss of weight following labor is, on an average, nearly the ninth part of the body-weight of a pregnant woman who has reached the end of the tenth month.

The loss of weight due to labor and childbed amounts, on an average, to the fifth part of the body-weight of the pregnant woman.

CHAPTER VI.

CHILDBIRTH.

IT has been well settled for many years that the pains of childbirth may be greatly mitigated, if not entirely prevented; and beyond even this, is the fact established, that in almost every country women are to be found who suffer no pain in childbirth.

It is reasonable to suppose that this exemption from pain could not occur in isolated cases, unless it were within the capabilities of the mass of the sex. It is within the personal knowledge of the writer, that healthy married women have gone to bed not at all apprehensive that the pains of childbirth were approaching, and yet within a few hours, and with only a few minutes' suffering and inconvenience, were happy mothers; having experienced no sickness, and lost no time in consequence of pregnancy, and recovering rapidly and easily after confinement. Drs. Dewees and Eberle, among the highest medical authority of this country, many years ago took the broad ground that pain in childbirth was a morbid symptom, the consequence of artificial, stimulated, or unnatural modes of life and treatment, and could be largely, if not entirely, avoided by appropriate habits and treatment. A suitable regimen will insure the pregnant female great tranquillity of mind and body, and little risk of injury to herself and child; whereas, the reverse will follow a free reign to appetite, indulgence to excess, or the use of improper articles of food.

Mrs. Gove, in her lectures to ladies, says:

" The functions of gestation and parturition are as natural as digestion. I know many mothers who, with their husbands, have adopted the ' Graham system '—or have given proper attention to diet, exercise, and bathing freely and constantly with pure cold water; and these mothers have abridged their sufferings in parturition from forty hours to *one hour*, and have escaped altogether the deathly sickness of the first three months of gestation. But they avoided all excesses as far as possible. We know that the Indians, the lower orders of Irish, and the slaves at the South, suffer very little in childbearing. Why is this? God made us all of one blood. Is it not that these, living in a less artificial manner, taking much exercise in the open air, and living temperately, have obeyed more of the laws of their being, and, consequently, do not suffer the penalty of violated laws, as do our victims of civilization ? "

RULES FOR PREGNANT WOMEN.

The following rules or suggestions are compiled from the writings of Drs. Gilman, Combe, Eberle,

Bell, Dewees, and other celebrated physicians, and are worthy of serious consideration :

Diet.—This should be light, not very nutritious, and rather laxative. For the first four months, let the patient eat vegetables, and especially fruits, freely ; and abstain from gross articles, from highly seasoned meats, and from stimulating drinks. After quickening, the digestion usually improves, and a rather more nutritious diet may be allowed. But as the term of gestation approaches, the diet should again be light; avoid, at this time, articles likely to produce flatulency. During the whole of pregnancy, the patient ought to pursue her usual avocations and mode of life, provided these conform to the suggestions above, and are compatible with the laws of health. *Longings* should not be indulged, or, if at all, quite sparingly ; they rarely occur to any extent " in a healthy woman of a well-constituted mind, but are peculiar to delicate, nervous, irritable, and, above all, *unemployed* women who have been accustomed to much indulgence, and have no wholesome subject of thought or occupation to fill up their time."

Influence of Atmosphere.—" Cold, rainy weather, and low, damp, miasmatic localities, have been recognized as disturbing pregnancy and causing abortion." Avoid the latter, and encourage sunlight, bright and cheerful rooms, and fresh air.

Exercise is the most valuable means of preserving the health of pregnant women. It should always

be taken in the open air, and carried so far as to produce fatigue, but not absolute exhaustion. Walking
is the best, but riding in an open carriage will do
well. Avoid horseback riding, unless the patient is
accustomed to it, rides well, and has a gentle horse.
The persistent sleeplessness with which some women
are troubled toward the close of pregnancy, is most
surely overcome or relieved by much exercise in the
open air, carried to fatigue ; this, with warm bath,
will do more than all the anodynes you can give.
Many delicate women, by confining themselves to the
house, and even to their chambers and beds, have
made themselves the victims of frequent miscarriages
—which might surely have been avoided by taking
regular out-door exercise, and attending to their domestic avocations. More harm is done by sudden
efforts, as in lifting, reaching after things overhead,
pulling, pushing, stepping with a bound so as to light
only on the forepart of the foot, or by jumping, than
by prolonged exercise, or even labor—though neither
of these last two is proper for women unaccustomed
to them.

Dress.—Any thing that compresses the body and
obstructs circulation does harm. Do not wear the
dress at all tight, nor low in the neck, so as to expose
the breasts to cold. " The custom of wearing tight-
ly-laced corsets, during gestation, can not be too
severely censured." The body and breasts must *not*
be forcibly compressed while nature is gradually
enlarging, both for the accommodation and develop-

ment of the fœtus. If corsets are worn at all, by all means loosen them fast enough, and sufficiently to admit of full breathing space. Every approach to absolute pressure should be scrupulously guarded against.

The Mind.—" Pregnant women should never be allowed to witness any scene that will be likely, very powerfully to excite, alarm or distress them ; the evil influence of rash impressions is well established. Even the more exciting pleasures of life they should partake of sparingly, as balls, parties, theatrical exhibitions, etc."

" An unkind word, a cold or severe look, or even apparent neglect, will frequently, in this state of health, derange the whole physical system, prostrate the most promising state of convalescence, and set medical skill at defiance. And because, in this morbid condition of the system, unkindness and neglect are more keenly felt, so, also, the kind offices of affection are doubly appreciated."

Summary.—" Regular daily exercise, cheerful occupation and society, moderate diet, pure air, early hours, clothing and mode of dress suitable to the season and to growth, and healthy activity of the skin, are all more essential than ever, because now the permanent welfare of another being is at stake, in addition to that of the mother. But any of these, carried to excess, may become a source of danger to both mother and child. Dancing, riding on horseback, traveling over rough roads, and vivid exertions of mind, have often brought on abortion. Cleanliness and fresh air are doubly necessary during gestation; hence, the propriety of a tepid bath every few days, especially in the case of females of the middling and higher classes, in whom the nervous system is unusually excitable. It promotes the healthy action of the skin, soothes the nervous excitement, prevents internal congestion, and is in every way conducive to health; but it must not be either too warm, too long continued, or taken too soon after meals.

DISEASES OF PREGNANCY.

The diseases of pregnancy vary with females according to their constitutions. Some are troubled greatly with sick headache, morning sickness, heartburn, costiveness, diarrhœa, piles, enlargement of the veins of the legs, swelling of the feet and legs, palpitation of the heart, fainting fits, toothache, salivation, cramps, etc., while others have none of the above troubles. It frequently happens that the sickly, weakly girl, on becoming pregnant, acquires *new* life and vigor from the changed circumstance of her condition, while the robust and well developed female will suffer seriously from it.

Morning sickness is one of the distressing affections of pregnancy, and lasts for about four months from the time of conception. This sickness is felt immediately after arising in the morning. The patient has no appetite for breakfast, and if any thing is forced into the stomach it is soon after expelled. After the lapse of three or four hours she feels quite well again, and the appetite for dinner can be gratified without fear of the return of the morning experience. This sickness arises from irritability of the stomach, and does not signify that the digestive organs are in a disordered condition, but rather a want of sympathy between the stomach and the newly-commenced being in the womb. Medicines in this case are seldom needed, as a mouthful or two of warm salt-and-water will in most cases relieve the sufferer.

Heartburn is another derangement the pregnant woman has to suffer during the earlier months of that condition; it is produced by acids forming on the stomach. The sufferer can nearly always find relief in a little magnesia or chalk. She should at all times keep her bowels well regulated, and thus avoid much of the suffering attending heartburn, as

Costiveness is one of the most common and troublesome of the diseases accompanying pregnancy, though it is not always the cause of heartburn.

Piles.—Of all the causes which operate in the production of piles, constipation is the most frequent. In pregnancy there is more of a disposition to costiveness than at any other time, and as piles are one of the results of this disordered function, so this disease is much more prevalent during the pregnant state. The diet in this case must be sparing in quantity and of mild quality, so as to leave after its digestion as little to pass through the bowels as possible. It is sometimes important in this case to take medicine to work on the bowels, say before retiring at night, thus giving it time to operate while the body is quiet; whereas by taking it in the daytime, while the patient is moving about, it frequently causes irritation and increases the suffering.

Swelling of the Limbs.—During the latter months of pregnancy the feet and legs frequently become much enlarged. This is owing to the pressure of the womb, generally being observed toward evening, about the ankles, and gradually ascending until the legs become of very large size. The female retires with legs much swollen, but toward morning the swelling reaches the head and diminishes in the legs, to return there again in the evening. This disease seldom needs any farther attention than the application of a piece of flannel, exciting the blood to action.

Fainting is of frequent occurrence during pregnancy or previous to the quickening. When it occurs simply sprinkle the face of the patient with cold water, and rub the extremities with the hands until a healthy circulation is established. Persons subject to such attacks should not take violent exercise, and should avoid warm rooms as much as possible.

Pregnancy being natural to married women, the easiest way to mitigate the sufferings attending is to

keep the system in good condition—the bowels regular and a little inclined to laxity, rather than cos-
tive ; diet wholesome ; have regular exercise, such as
walking in the open air, not, however, to such an
extent as to weary ; avoid tight lacing, and keep the
mind employed, so as to avoid brooding over imagi-
nary troubles. It is natural for the pregnant female
to feel gloomy, and every possible means should be
used to promote cheerfulness. If these rules be followed strictly, much suffering may be prevented.

DEATH-RATE IN CHILDBIRTH.

Because of the erroneous impressions prevailing
among women as to the great danger of death in
child-bearing, we have gathered the following statistics from the most reliable sources. These embrace
the large number of 18,285 cases of labor with only
305 deaths, or at the rate of a little over sixteen
(16.7) to the thousand. But these nearly all occurred
in hospitals, where, notwithstanding the comforts
and conveniences supplied, the soothing and affectionate attentions of home were wanting.

Dr. H. Corson, of Montgomery County, Pennsylvania, in 2,387 cases of consecutive labors, in his
own practice, lost but nine by death in child-bed—
or *less than four cases in a thousand.* None of these
were in hospitals, but in their own homes.

In the year 1865, in London, there were in 30
workhouses 1,754 cases of childbirth, without a

7

single death of the mother; and in nine other workhouses there were only 16 deaths in 974 cases of
childbirth—an average death-rate of 6 per 1,000
cases.

In Queen Charlotte's Hospital, which possesses
a world-wide reputation as a school of obstetric
practice, there were 2,268 cases of childbirths and
90 deaths in the seven years from 1857 to 1863—
a death-rate of 40 per 1,000 cases; whereas, in the
British Lying-in Hospital, in the thirteen years
from 1848 to 1861, the death-rate was only 7 to the
1,000, or 11 deaths in 1,581 cases. This death-rate
was still lower, only 3.5 per 1000 in the eight years
from 1856 to 1863, in the out-door midwifery department of St. George's Hospital, or 10 deaths out
of 2,800 cases.

In the Rotunda Hospital, at Dublin, Ireland, in
five years, from 1857 to 1861, there were 169 deaths
out of 6,521 cases of childbirth—a death-rate of 26 to
the 1,000.

Several causes are assigned for the extraordinary
mortality at Queen Charlotte's Hospital, foremost of
which is the great number of single women received
there; the mortality of the unmarried being far beyond that of the married women in childbirth.
During thirty-five years, in that institution, the
death-rate of the latter is only 18, while that of the
single women in childbirth is 35 per 1,000 cases.
Amongst unmarried women, a great mortality has
always prevailed on their passage through the puer-

peral stage. Dr. Bradie says—and it speaks strongly for home influences—that " women delivered at their own habitations, though often living in the greatest filth and poverty, with only one room to accommodate the wants of a whole family, do infinitely better than those who are removed to a spacious, well-ventilated building, with every comfort and attention that can be desired."

CONTINUANCE OF LIFE OF THE FŒTUS AFTER THE MOTHER'S DEATH.

Prof. Breslau, after tabulating twenty experiments and close study and observation, concludes that the human fœtus always survives the mother if her death is rapid and violent, as from bleeding, blows on the head, apoplexy, etc. The Cæsarean section should be performed as quickly as possible, but unless within fifteen or twenty minutes after the mother's death, there would be no hope of living or of an " apparently dead " child.

In four cases, by Prof. Martin, within ten minutes to half an hour, none were born alive, and in one hundred and forty-seven cases, collected in Caspar's *Wochenschrift*, only three instances of living children occurred.

CHAPTER VII.

EFFECTS OF THE IMAGINATION ON UNBORN CHILDREN.

TO the newly married woman and prospective mother, we can utter no more friendly and perhaps more timely word of warning than this—preserve a quiet and cheerful state of mind, and avoid violent exercise or straining of the body, during pregnancy. The female imagination, at this time of strange and nervous feelings, is unusually active and sensitive. The brain, temperament, and physique of the unborn child is as impressible as a mirror, or as the plate in a camera. It is the experience of all observers, and consistent with the established laws of physiology, that parents transmit to their offspring their own mental and personal peculiarities. It is no less true, that the strong mental emotions of the mother—especially those of terror and disgust—are pictured upon the persons or perpetuated in the minds and dispositions of the children.

To enforce the warning above—to present the strongest possible inducement to the pregnant woman to guard herself against all undue excitements—we present the following instances of body-marks upon children, of a singular character; some of which are within our own knowledge, but most of them gathered from authentic medical sources.

BODY-MARKS.

A distinguished Presbyterian divine of Indiana, now fifty years of age, has upon, or just below, the right shoulder-blade a distinct representation, or *fac simile* of an ordinary coffee-grain, or bean, which stands out, or projects from the flesh, has uppermost the exact crease of the inner, or flat part of the bean, and is of the color of browned coffee. The mother, before his birth, visited a neighbor who was engaged in roasting coffee, in the pioneer days of life on the border, and was seized with an intense longing for that beverage, of which she had been deprived for months. On inhaling the aroma, she uttered an exclamation of surprise and delight.

A lady living not far from Cincinnati, Ohio, now over forty years of age, has upon her right leg, above the knee, the very marked outline and color of a rat— the result of a severe fright to her mother, from an animal of that species climbing under her clothes, while trying to escape from a dog.

In Covington, Ky., Mrs. A. S., in the year 1866, gave birth to a child singularly deformed about the lower part of the face, nearly half of the left jaw having no flesh to cover it, an ugly discoloration as if produced by a gun-shot wound, and at its every breath emitting a horrible and fetid stench. It did not live a week. The mother, during her pregnancy, had watched over, and kindly nursed for several weeks, a young man who was similarly disfigured and loathsome, from a wound received in battle.

The Albany (N. Y.) *Knickerbocker*, June, 1866, mentions the case of a young man named Abriel, a returned soldier, on First Street, in that city. When he came home he was suffering from a Minie-ball wound through the fleshy part of his right arm. It became so bad that the attending physician talked seriously of amputation. This worked seriously on the mind of the young wife (he had but a short time previously got married). She cared for and dressed the arm regularly, and paid every attention to it, not wishing to see her husband with only one arm. This was eight or nine months ago. The wound got well, and the arm was saved. The other day the wife of Mr. Abriel gave birth to a child who had one developed arm, but the other was a stump similar to one which the poor wife's mind was impressed with at the time the surgeons were talking of taking off her husband's. Amputation could not have produced a more beautiful stump; and what is more, the scar of the bullet hole, so visible on the father's arm, was as visible on the child's arm at the base of the stump, as if really inflicted by a ball. This is the most remarkable case of "child-mark" ever known. It has attracted the attention of all our leading physi-

cians and surgeons. The child is a healthy and
beautiful one, perfect in every respect, save the arm
referred to.

The following case is given upon the authority of
Dr. S. P. Crawford, in the Nashville (Tenn.) *Journal
of Medicine and Surgery.* A Mrs. James, of Washington County, Tenn., was burned to death by the
explosion of a kerosene oil-can. The face, legs, arms,
and abdomen were completely vesiccated, and the skin
in many cases was entirely destroyed. She was in the
last stage of gestation, and lived only twelve hours
after the accident. The movements of the child were
distinctly felt three or four hours after the accident.
A short time before the death of the mother she gave
birth to a child. The child was at full maturity, but
still-born. It bore the marks of the fire corresponding to that of the mother. Its legs, arms, and abdomen were correspondingly vesiccated, having all
the appearance of a recent burn.

The Hopkinsville (Ky.) *Constitution,* of June
20, 1866, states, on respectable authority, that a lady
residing at Fairview, in that county, gave birth, some
six weeks ago, to a child that was a most singular
compound of man and reptile. The lower portions
of the child are natural in their formation, and partake of the characteristics of the *genus homo,* but the
body and the head are similar to the body and the
head of the rattlesnake. The mother, every time she
is compelled to give nourishment to the child, is
thrown into convulsions. The singular formation of

this creature is thus accounted for : Some time during pregnancy, a rattlesnake attacked and greatly
frightened the mother of this creature, but fortunately,
however, did not injure her.   At the date of writing
the child is still alive.   Its parents are among the
most respectable people of Christian County.

A case is recorded of a child having large scars at
birth, directly across the knee-pan of each knee. They
were of irregular form, and looked as though they had
been made with a nail.   There was no doubt but that
these scars had been made while the child was still in
the womb.   On inquiry it was found that the mother
had spent several hours each day for several days, on
her knees, at the side of the cradle of a sick child.

A lady of very even temper and delicate constitution, during the second month of gestation, was presented by her husband with a pair of ear-rings. She
was exceedingly desirous to wear them to an evening
party, but after persistent effort found it impossible to
insert the hoop into one ear, as the hole had partially
closed up.   The attempt was abandoned, with some
disappointment.   When the child was born it had a
hole in the center of one lobe of the ear, and a deep
cleft running downward for a quarter of an inch.

Mr. A., of northern New York, married. His
pecuniary circumstances at that time made offspring
undesirable.   Within a year, however, it became
evident to the wife that their wishes were not to be
longer realized. She made known her situation to
her husband, and was terribly shocked at the dis-

satisfaction with which he received the news. He left the house and was absent about an hour; returning, he found his wife in tears. Surmising the cause, he immediately told her that he was rejoiced to learn of her situation, as he was satisfied with his pecuniary affairs, and felt that offspring would be desirable. The wife dried her tears, but had a constantly-recurring presentiment that their child would bear some mark from her sorrow. The relief of the parties was great at the birth of a healthy, well-formed boy. After a few months this child would scream at the approach of the father; and no kind words, affection, or promises could ever lead it to approach him. The son grew to manhood, and when last heard from was a rising member of the bar. He could never get over this natural aversion to his father, much as he desired to, and made most painful efforts to accomplish it. The most singular and painful circumstance was that, though eloquent to others, he had never been able to speak a word to his father.

A lady was in constant attendance upon her dying father; his disease was a cancer on the forehead, and required daily dressing; this was done by the daughter, who was then in the early period of pregnancy. In a few months the father died; and the daughter was delivered, at the full period, of an infant with a large tumor on its forehead, precisely like the grandfather's cancer, and which soon caused the child's death.

A gentleman and his wife visited a zoological garden in New York. They were startled by the ferocity of a Bengal tiger, at being disturbed from sleep. In time she gave birth to a healthy boy, who, as he grew older, on being crossed by any of his playfellows, would growl, and scream, and fly at them with all the ferocity of a wild cat. As he advanced in years, it became necessary to have an older person accompany him, to prevent injury to his playmates. The screams or cries were just like those of cats when fighting. This happened before he was five years old, since when he has not been heard from.

A farmer desired his wife, to whom he had been married but a short time, to assist him in slaughtering a calf. The animal was thrown upon its side, and when the man was in the act of applying the knife to the throat, it sprang to its feet, receiving at the same time a severe cut across the mouth and severing one of its ears. In due time the woman was confined, and gave birth to a child with a hare-lip, and each lip deeply cut through, and with only one ear. The child died while the physician was in the act of closing up the fissure.

A man addicted to strong drink married a woman whose favorite pet was a cat. The husband, seeing her fondling it, became enraged, and in a fit of drunken passion dashed its head against the hearth. Their first child, born a few months after, bore strong resemblance to the cat, in the expression of the face, head, and eyes, with long and depressed nose, and short fingers with sharp nails, while the limbs were deformed. This child was passionately fond of the mother; but shunned the father. Up to nine years of age, it had not learned to talk, but made known its wants in a sharp, painful, cat-like yalling.

At a wedding party given to a young couple soon after their marriage, the bride, from some cause, received the ·contents of a wine-glass on her face, neck, and shoulders, which caused her some confusion, and rather violent anger. Their first child came into the world with its face, neck, and breast well covered with claret—the mark or color becoming brighter as the child grew in years.

The wife of a farmer attempted to extract a favorite heifer from the mire, but being unable to accomplish it alone, she sat down and commenced patting the animal with her hand, and toying with a little curl in its forehead. Some months afterward a child was born, and grew to manhood, having a little tuft of hair in its forehead very unlike the hair on its head, and resembling in redness, coarseness and texture that of the animal.

A few years since a Baptist clergyman was riding with his wife, in a covered sleigh. They met another team, the horses of which were running away. In passing, the head of one of the horses came in contact with the cover of the sleigh and carried away a part of it, slightly injuring the gentleman and frightening the lady terribly. She imagined that the collision had taken the top of her husband's head off, and could not be satisfied until she repeatedly put up her hands to find out for certain. Some months after, she gave birth to a child with a face, but with no forehead, head or brain—the hair starting out immediately above the eyes. It did not live long.

The writer is reminded of the case of a sow in Greensburg, Indiana, which was badly frightened at the sight of an elephant, belonging to a caravan then exhibiting there. In a few weeks this sow gave birth to a healthy litter of pigs, among them a perfect little elephant, with tusks, ears, trunk, feet, eyes, and skin, the image of what had caused her terror. This was preserved in alcohol, and twelve years later could be seen at any time in a drug store in the town.

Whether the impressions of the mother's mind do or do not influence its formation, we can only judge by facts. How this may occur we know not, but " there are more things in heaven and earth, than are dreamed of in our philosophy."

CHAPTER VIII.

MARRIAGE OF KINDRED.

MARRIAGES between cousins or other blood relations, it is almost universally believed among civilized people, deteriorate the race—engendering beings afflicted with some vice of conformation, or with degeneracy, such as mental imbecility or positive madness, epilepsy, idiocy, and cretinism, deafness and dumbness, or both united, the whiteness of the Albinos, sterility, etc. As a general proposition this is certainly true, but is modified or undeveloped—if actually existing—in many cases where the intermarrying relatives are widely varying in their temperaments.

Dr. Mitchell, of Edinburgh, Scotland, in February, 1862, read before the Medico-Chirurgical Society, of that city, a very able and quite exhaustive paper upon this subject—arriving at the following among other conclusions:

1. It is a law of nature that offspring is injured by consanguinity in the parentage.

6. That where the children of such seem to escape,

the injury frequently shows itself in the grandchildren.

7. Idiocy and imbecility are the most frequent forms of *mental* disease resulting.

9. Transmissible peculiarities of all kinds are apt to be intensified, and diseases are aggravated.

Out of seven hundred and eleven idiots examined in his researches, one out of eight was the fruit of consanguineous marriage, and of those born in wedlock, 1 in every 5.8, or more than every sixth idiot was found to be the child of cousins.

In Great Britain 1 deaf-mute in 16.7 was found to be the child of cousins.

Dr. Peet, principal of the New York Institution for the Instruction of the Deaf and Dumb, has published statistics from which he deduces the striking conclusion that in Europe, generally, the chances of the birth of a deaf-mute child are twice as great as in the United States, or 615 in a million there against 278 in a million here. One of the most prominent causes he assigns for congenital deaf-dumbness is the intermarriage of blood relations.

As to the effect of such alliances in producing blindness in the offspring, Dr. S. G. Howe, principal of the Institution for the Blind, at Boston, one of the most active philanthropists in America, says: " There can be no mistake at all about the fact that the tendency to have defective offspring is greater where parents are defective, than with others. But here is a point that leads people into error. It does not follow because a person is defective in his hearing the defect will take that form in his offspring; it may strike somewhere else. The child may be defective in physical strength or mental capacity. But there is the defective germ, and it will manifest itself. It may skip one generation and manifest itself in the next. I know of thirteen blind children, in a neighboring county, the descendants of one blind man who married his cousin. In the first generation there were no blind children. You would look around and see these children all happy, all enjoying the blessings of sight, and say, 'It is all moonshine, this idea about defective people marrying.' In the second and third generations came thirteen blind children (from the intermarriage of a blind man with his cousin). I think six of these have been in our institution."

A report, read before the National Medical Association at Washington, by Dr. S. M. Bemiss, in 1858, shows that over ten per cent. of the blind, and nearly fifteen per cent. of the idiotic in the different State institutions, were the offspring of kindred parents.

The Report of the Commissioners of the Kentucky Institution for the Education and Training of Imbeciles or Feeble-minded Children, in a passage urging the prohibition of first-cousin marriages by legal statute, uses the following language: " We deem it our duty to the interests of humanity, as well as to the pecuniary interest of the State, to bear our testimony in addition to the abundant statistics hereto-

fore collected and published by physicians and philanthropists, and to the observation of every close
observer, as well as to general considerations of
propriety, that a large percentage of deaf mutes and
of the blind, a limited percentage of lunatics, and,
no doubt, a much larger one than either of feeble-
minded or idiotic children, are the offspring of the
marriage of *first* cousins. Our charitable institutions
are filled with children whose parents are so related
—sometimes as many as four from one family; and
we have known, in the case of idiots, of a still larger
number in a family. It is a fearful penalty to
which persons so related render themselves liable by
forming the matrimonial relation, and which they,
in nearly every instance, incur, not indeed in all, but
in one or more of their offspring. Instances, we do
not deny, may be shown where a portion of the children—one or more—may inherit from both parents,
where possessed of high mental and bodily endowments of a common origin, enhanced and remarkable
qualities of body and mind; but it is generally at
the expense of unfortunate and deeply afflicted
brothers and sisters. We believe few instances can
be given where such enhanced endowments are common to *all* the offspring; while instances are not unfrequent where nearly all, and, in a few, perhaps,
every child is afflicted either in body or mind, and
sometimes in both."

Dr. Carpenter, of the University of London, in his
" Principles of Human Physiology," uses the follow-

ing strong language: "The intensification which almost any kind of perversion of nutrition derives from being common to *both* parents, is most remarkably evinced by the lamentable results which too frequently accrue from the marriage of individuals nearly related to each other and partaking of the same 'taint.' Out of 359 idiots the condition of whose progenitors could be ascertained, 17 were *known* to have been the children of parents nearly related by blood, and this relationship was *suspected* to have existed in several other cases in which positive information could not be obtained. On examining into the history of the 17 families to which these individuals belonged, it was found that they had consisted in all of 95 children; that of these no fewer than 44 were idiotic, 12 others were scrofulous and puny, 1 was deaf, and 1 was a dwarf. In some of these families all the children were either idiotic or very scrofulous and puny; in one family of 8 children, 5 were idiotic."

According to "*Chambers' Encyclopedia,*" the result of an examination into the congenital influences affecting deaf and dumb children in Scotland, was that of 235, whose parentage could be traced, 70, or nearly 30 per cent., were the offspring of the intermarriage of blood relations. The physical deformity and mental debasement of the Cagots of the Pyrenees, of the Marrons of Auvergne, of the Sarrasins of Dauphiné, of the Cretins of the Alps, and the gradual deterioration of the slave population

8

of America, have been attributed to the consanguine-ous alliances which are unavoidable among these unfortunate people.

A medical writer of some ability, who has had the opportunity of observing the effect of the intermar-riage of near relatives in the Eastern States—though, of course, such marriages for several successive gen-erations, will produce more marked results than a single infringement of the physiological law—gives the following cases :

Case 1.--W. F., in a New England city, married his niece, J. E. They had two children, a boy and girl. The boy died at the age of twelve, an idiot. The girl grew up to woman-hood, but from scrofulous disease of the vocal chords was never able to speak except in a whisper, and though possessing fair average mental endowments, was deformed in body, and died at the age of twenty-three, of tuberculous phthisis, after severe and protracted suffering. There was neither scrofula nor phthisis in the preceding generations.

Case 2.—J. H., originally from New England, but subse-quently a resident of New York, married his niece. Though of different temperaments, their child (they had but one living one) was imbecile in intellect, and of feeble and scrofulous con-stitution. After some years the wife died, and H. married again, this time where there was no blood relationship, and a woman of temperament strongly in contrast with his own. By this wife he had five healthy and beautiful children.

Case 3.—G., from Vermont, married his cousin; the husband was bilious, sanguine-lymphatic, the wife bilious encephalic. They had three children, all daughters, all of scrofulous ten-dency, one deaf, the others, though with fair mental endow-ments, yet of intensely nervous and excitable temperaments.

Case 4.—G. W. married his cousin, both being of sanguine-lymphatic temperament. They had three children, two of them sons; the eldest was an idiot from birth; the second was of fair intellect, but scrofulous and consumptive, dying of phthisis at the age of thirty-two. The daughter was of sound mind, but suffered from scrofulous disease of the hip-joint, which made her a cripple. No scrofula in the family before.

CASE 5.—T. P., of E. Conn., married his double cousin, H. P. They were of different temperaments. They had five children. The eldest, a son, was weak-minded—half-witted, as the phrase was among the country people. The second son was a hopeless, mischievous imbecile. The third son was not deficient in intellect, but a great sufferer from scrofulous disease of the joints. The fourth child, a daughter, was of weak intellect. The fifth, a son, died of disease of the brain (tubercular meningitis, I believe) in his eighth year.

A distinguished French physician, Dr. Dévay, after a lengthened inquiry into the subject, concludes that " we may in future include consanguinity in the catalogue of morbid etiology, as far as regards the human race." With animals, also, the evils are equally manifest.

Dr. J. L. H. Down, assistant physician to the London Hospital, in 1867, in an able paper, attempted to refute the prevalent opinion above referred to, that marriages of consanguinity are uniformly productive of evil to a greater or less extent. His observations were quite extensive, embracing eleven hundred and thirty-eight cases of idiots, and he certainly establishes that there are other causes which are frequently and sometimes largely influential in producing the evils of such marriages. His reasoning is as conclusive as that of the Scotch peasant, on his native heath, who, in reply to a suggestion that it is always raining or misting there, several times repeated notwithstanding his broad " nae, nae," at last satisfied the inquiring American by still more broadly saying, " it does not always *rain* here ; sometimes it *snaws.*" It is not consanguinity in marriage, *alone* and invariably, that produces bad results, but other causes sometimes combine to produce, if possible, more disastrous or painful effects.

The most plausible advocacy of the intermarriage of relations is by a French physician, Dr. Auguste Voisin, who, in the year 1864, passed a month in the borough of Batz (Lower Loire), " the inhabitants of which," he states, " have, for several centuries past, been in the habit of contracting consanguineous marriages, and live almost isolated from the surrounding populations, intercourse with whom they pretend to despise." He examined the results of the marriages between blood relations, which are at present to be found there forty-six in number. The antecedents of the husband and wife have been inquired into; they have been themselves questioned, as well as their children, and examined intellectually and physically. M. Voisin has, besides, sought information from the mayor, the curate, and the elders of the country. From all this he has drawn up statistics, the principal results of which are as follows :

'Out of forty-six households, of the borough of Batz, who are united by ties of blood, five are composed of second cousins; thirty are between the offspring of second cousins, and ten between cousins of the fourth degree. The five marriages between second cousins have produced twenty-three children, of whom none were infirm or malformed at their birth; two died of sickness. The thirty-one marriages between the issues of second cousins have produced a hundred and thirty children, none of whom have any infirmity or congenital affection; twenty-four have died of acute diseases. The ten marriages between cousins of the fourth remove have given .twenty-nine children, three of whom have died of acute diseases; the others are all well. The health of the fathers and mothers of these married couples is good and exempt from all chronic maladies; that also of the individuals themselves and of their children is excellent. They are generally tall, and the configuration of the head with most of them corresponds to a uniform type.

" Out of forty-six couple thus studied, only two are sterile, where the bride and bridegroom are relatives of the third degree. The forty-five others have produced in all a hundred and seventy-four children, out of whom twenty-nine are dead. Two striking facts result from these figures: the fertility of marriages between blood relatives of the borough of Batz is greater than the average fertility of marriages in France, and the mortality of children is relatively less.

" M. Voisin sums up his observations that consanguineous marriages, in the borough, have produced no maladies, no degeneracy, and no malformations; and, in spite of its secular custom, the race has remained very fine and very pure. This should not surprise us; on the contrary, ' I think,' says the author, ' that this result is to be attributed to *the exceptional topographical conditions of the country and its climate, to hygiene, to habits, to the morality of the inhabitants, and the absence of all morbid inheritance.'* "

From all these facts, M. Voisin concludes that " *under the said conditions of good selection,* consanguinity does not injuriously affect the offspring and race, but, on the contrary, improves its good qualities."

DR. POWELL'S PHYSIOLOGICAL INCEST.

Dr. Wm. Byrd Powell, an eccentric and talented physiologist and medical philosopher, of the Eclectic School, died in Covington, Kentucky, in May, 1866. In his last will and testament, was the singular direction that his head should be dissevered from his body and delivered to one of his enthusiastic students—which unnatural legacy, upon the personal application of the legatee to the court, and by the latter's authority, was literally executed. Dr. Powell was the originator of some strange but striking theories in reference to the laws of marriage, which were singularly illustrated in several cases to which his attention had been attracted. Marriages in and in, or in consanguinity, to an unascertained extent, degenerate the species; whereas violations of the "laws of the human temperaments," he claims, are more positively visited with serious results—either entire sterility or the early disease and death of the offspring. His main theory seems to be, that young people who closely resemble each other in complexion, shape of person, and form of head, must not intermarry. A young man intending to marry will be less likely to make a mistake if he select as his companion one who least resembles him in personal appearance. The offspring of

marriage is usually expected to develop the prominent qualities of the parents. If these qualities are
mainly the same in both parents, there is likely to
be an unhealthy or disproportionate, perhaps dangerous, development of them in the child. But,
warning our readers to be careful to obtain a correct
idea of his theory—by closely analyzing the tech-
nichal language he employs—we prefer to make
literal quotations from his latest writings, published
in the summer of 1867, after his death :

"Some fifteen years ago I was induced to suspect that a
physiological incompatibility obtained between the sexes in
relation to progeny, and it induced me to adopt an active and
careful course of observation of progenitors and progeny. The
result has been an entire conviction of the truth of my suspicion. This incompatibility prevails extensively in society,
and is, I am confident, the cause of all the idiocy, much of the
insanity, all of the tubercular consumption of the lungs, and
of mesenteric glands of the abdomen, and of all the scrofulous
forms of disease incidental to the human race. This conviction
induced me to denominate this incompatibility PHYSIOLOGICAL
INCEST, *by which I mean, physiologically unlawful marriage.* ·

"This physiological incest obtains thus: Certain of the human temperaments are incompatible in relation to progeny, and
this view of the fact has, *a priori*, a much greater probability
of truth in it than the assumed cause, consanguinity; because
temperament is a positive condition, and, therefore, capable of
being an agent; but consanguinity is merely a relation, and,
therefore, incapable of agency. After a few explanatory remarks in relation to the temperaments, I will show how this
incest obtains."

THE FOUR TEMPERAMENTS.

" I adopt that system of the temperaments, with some modification, which has descended to us from the ancients, the one
with which the public mind is most familiar. It comprises
four temperaments, viz.: sanguine, bilious, lymphatic, and
melancholic. The description given of the last so strongly indicates a diseased condition, and the fact that all the illustrations I ever saw of it being diseased subjects, caused me to dis-
card it from the catalogue of temperaments; and I did the
same with the so-called nervous temperament, and for the
same reason; nevertheless, early in my investigation of the
subject, I became convinced that humanity was distinguished
by four temperaments; and, after close and long-continued observation, I discovered what I deem to be the fourth temperament; and thirty years of additional observation have only
confirmed me in the conviction that I did discover the fourth
temperament, and I denominated it the encephalic.

"A few words of explanation now become necessary. The
sanguine and bilious temperaments constitute a *sine qua non* in
the transmission of life and the perpetuity of the species. I
infer, therefore, that they were originally founded in the constitution of the race, and hence I denominate them the vital
temperaments. The lymphatic and encephalic temperaments
I hold to be acquired conditions—to have resulted from influences incidental to civilization. They can not transmit life,
nor are they essential to life; hence I denominate them
adjunctive temperaments. The two following laws, which
have been deduced from an immense number of observations
during a period of fifteen years, explain the rest:

" 1. The marriage of a person with another of the same temperament is incestuous.

" 2. When an adjunctive temperament enters into the constitution of both progenitors or parties to a marriage, it will
be incestuous.

"These two laws are sufficient to enable those who understand the temperaments to distinguish accurately all incestuous parties. Nevertheless, I will add a case or two that will illustrate both of them:

" I. When both of the parties to a marriage have the sanguine encephalic temperament, their children will die young of dropsy of the brain, or of tubercular inflammation of its membranes.

"II. When both of the parties to a marriage have the bilious encephalic temperament, their children will be idiotic.

" III. When both of the parties to a marriage have the bilious encephalic lymphatic temperament, their children, in the proportions of 5 to 7 or 9 to 11, will be dead-born, and the others will not live two years, respectively. These three cases illustrate both laws.

" I will now illustrate the second law exclusively:

"IV. When one party is bilious lymphatic, and the other is sanguine bilious encephalic, their children will *all* die young of tubercular consumption of the lungs or abdominal glands, although neither of these forms of disease was ever in the ancestry of either party.

"V. If one party be sanguine encephalic, and the other bilious lymphatic, the progeny will die young of tuberculous forms of disease.

" VI. If one party be bilious encephalic, and the other sanguine encephalic bilious lymphatic, the children will, sooner or later, become insane.

" In the three preceding illustrations the temperaments of the parties, respectively, are greatly different, yet, as in each case, both parties partake constitutionally of an adjunctive temperament, all the marriages are respectively incestuous.

" A crude notion of the first law prevails extensively in this country; hence, we hear it frequently said that a person, in seeking a companion in marriage, should select one who is as opposite to himself or herself as possible; hence, a bilious lymphatic, who has a full and rotund body, dark brown or black hair and eyes, with a dark, sallow complexion, sets out to look for a wife, and finds a lady of a lean, spare person, and so gaunt about the abdomen as to appear disemboweled, with light hair, fair skin, and blue eyes. He is delighted in having found the right one, as he believes, because she is the opposite of himself as it is possible for one to be; he proposes marriage to her, and she, being influenced by the same notion, accepts of him, and then in six or seven years, or less time, they are greatly surprised that they can not have a child to live five years. These parties constitute case 5. There is an adjunctive temperament in both of them. This blunder is often made. *I know of no means by which incestuous marriages can be avoided, except a knowledge of the temperaments.* Ninety-five times in a hundred, young people in observing the first law violate the second.'

"*Law* 3. In all marriages with a view to, or expectation of, progeny, one of the parties should have an exclusively vital temperament, and then it is best that the other party should be more or less adjunctive.

"1. Illustration: one party sanguine, the other bilious, encephalic, or lymphatic, or bilio-encephalic lymphatic.

"2. One party bilious, the other sanguine encephalic, or lymphatic, or sanguino-encephalic lymphatic.

"3. One party sanguine bilious, the other sanguine lymphatic or bilious lymphatic, sanguine or bilious-encephalic, or bilio-encephalic lymphatic.

'These are all good marriages.

LAWS OF THE HUMAN TEMPERAMENTS.

In a subsequent article, upon "The Laws of the Human Temperaments, in relation to Marriage, and the consequences of their violation—illustrated by cases in practice," Dr. Powell again defines his position with regard to the human temperaments, to insure his being understood, thus:

"I define temperament to be a mode of being, *sui generis*, compatible with life, health, and longevity; and the elementary temperaments to be the Sanguine, the Bilious, Lymphatic, and Encephalic. The views usually entertained of the three former will answer my present purpose ; and the fourth is that in which the cerebrum is largely developed, with a reversed condition of every other part of the system. Its most distinguishing peculiarity is an expansion of the cerebral hemispheres throughout the full extent of the parietal ridges.

" Although, from time immemorial, it has been known that marrying in and in, or in consanguinity, was attended with a degeneracy of the species; yet, to the extent of my reading, it has not been known, that even stronger and more fatal incompatibilities to marriage alliance obtain outside of this circle, and with the best of health in the parties, and, in the abstract, the best of physical and mental organizations.

"In 1837, in the State of Mississippi, I took boarding in a private family, where I was never more pleased than I was with my host and his wife; both of them possessed vigorous health, both had the best of mental endowments, both were of rather a full habit of body—being of the same temperament, sanguine bilious lymphatic. I anticipated seeing, when they should return from school, some of the finest models of humanity in their children. But I was greatly disappointed. I inquired of the lady how many children she had. She answered six, that her three eldest were dead, and she greatly feared that the remaining three would go in the same way (I thought it would be well if they did); but the wherefore was a great mystery to her, as the family of her husband, like her own, had always been remarkable for their generally good health. At that time, the fact was as much of a mystery to me as it was to her. The three remaining children had large and badly-formed heads, small necks, chests, muscles, and bones. From their organization, I concluded that her fears that they would go as the others had gone, were well founded.

"My study of the human temperaments commenced eleven years before, or after I became disgusted with all that I could find in books or hear in schools about them ; but I had not observed any such family results before, and at this time did not regard them as holding any necessary relation to them. From this time forward, I observed closely every family I met with.

" It was not long before I met with an encephalo-bilious man and wife, whose two elder children, about ten and twelve years old, respectively, were so rachitic that they had never walked. They had three children; the youngest, a little girl, who closely resembled her father, possessed some promise.

" In a few weeks more I met with another encephalo-bilious man, whose wife was sanguine encephalo-bilious, with the two latter elements preponderating. One of their children was too idiotic to be trusted with business; another one, a married woman, had consumption; and the two remaining ones, though not wanting in intellect, were so imperfectly constituted that they could not probably live beyond maturity. The husband did not think that consumption had ever been in either branch of the family.

" In six months I saw enough to induce me to believe that the temperaments had something to do in the premises, and I began to treat of it in my lectures, and cautioned young people against uniting themselves with those who possessed a similar complexion and make of person.

" At Raleigh, Tennessee, I lectured one night, and treated this subject at considerable length. Next morning a fine looking and intelligent planter called upon me, and, after some desultory conversation, told me that he had had the pleasure of hearing my lecture the night before; and added that his wife resembled him very closely, having the same colored hair, eyes, skin, and make of person, and that both of them had always enjoyed very good health. 'Now,' said, he 'I desire your opinion as to the prospects of our children.' I answered, 'It is possible that you have had none; and if you have, I think it very probable that they will not live to be ten years old.' He replied, 'Your latter opinion is correct; my wife had six, all of whom are dead, and neither of them lived to be ten years old. But as my young negroes did well, I could not suppose my residence to be the cause; and hence, I was unable to comprehend why we should lose our children, till I heard you last night.' The old gentleman informed me that the mother of those children died of cholera, and that he had married again. His present wife was lean, and had black hair and eyes, and a dark skin. 'Now,' he inquired, 'what is my prospect for children?' I assured him that it was as good as he could desire. With evident emotions of pleasure, he remarked, ' We have three fine boys.'

" Several months elapsed before I could give the *rationale* of such results. I had long before come to the conclusion that nature, or the procreative law, was solicitous for the good and safety of the species, even at the cost of individuals, but how the good of species required the preceding sacrifices, I could not imagine; on the contrary, I thought that such parents were such as nature should desire to perpetuate the species. At length I met with a man and his wife, in both of whom the phrenological region of reverence and hope was very high developed, while the head of each child was as flat on top, or nearly so, from one parietal ridge to the other, as the table I am writing on. The reflection of a few minutes satisfied me why this should be so. Nature aims at the good of the species; and if the combined or united strength of this portion of the brain in the parents should be entailed upon the children, and these children should entail it in the same way, nothing less than monstrosity could be the ultimate result. Even the first generation would be unfit for usefulness, except as mere dependents. Thus I became led to a conclusion with reference to the previous cases. It would have been exceedingly unfortunate for the first and last of those parents to have entailed the united strength of their vital vigor, or cellular repletion. Nature has forbidden it, and therefore she will not allow the fruits of such disobedience to live, if produced. But more of this, by and by.

"I am now acquainted with a family in which the husband is sanguine bilious lymphatic, and the wife bilious encephalo-lymphatic; both are much advanced in years, and in good health. Four of their children have died of consumption, another has now the disease, and the sixth has an organization strongly indicative of a liability to the same.

"I have a legal acquaintance of the quadruple constitution, who invited me, in 1853, to take tea with him. Soon after my arrival he presented his little son to me. I discovered that he would not probably live to maturity; and directed my attention to the mother, and observed that she had the same combination of the temperaments, though not in such balanced proportions, as the husband. The father asked my opinion of the child. I told him that he would not probably raise him—that his brain was too much developed for his body. He remarked that he had lost four children of brain-fever; that all of them had much the same organization; but that this one was more stout and healthy, and progressed at school with great rapidity. I think this boy is still living.

"A sanguine-encephalic Judge, in Alabama, invited me to dine with him. His little son came in from school, when the father remarked that he had a prodigy in that son, to which I replied, "It is a great misfortune." He evidently expected me to express some admiration of the child, and in his confusion, asked me why I thought it a misfortune. At this moment his wife came in and was presented to me; whose constitution I perceived to be sanguine encephalo-bilious—the two former elements preponderating. He informed her of what I had said, which occasioned in her some alarm, and she expressed, as the Judge had done, a desire to know my reason. I told them that without a vigorous circulation, digestion, and assimilation, the brain could be of but little use; and assured them that the child's arterial, pulmonic and digestive systems were not-sufficient to sustain so large a brain. I advised them to give but little attention to his scholastic education—to give him a free use of the play-grounds, and to send him as much as possible to a dancing school. In about three years after this event I called upon the Judge; his wife met me with tears in her eyes, and exclaiming, "My poor boy is dead, and I believe that if we had taken your advice he might have lived."

"From what I could learn, it seems there had been a convention of the school teacher, the parson, and the parents, to determine in relation to the course I had prescribed for the boy. The parson had some doubts of phrenology, and admitting it to be true, he contended that I might make mistakes in the application of it; and as the boy progressed rapidly at school and had excellent health, it was decided to keep him at school. The mother, in this convention, constituted the minority—she favored the adoption of my course, but this was referred, by the majority, to her maternal fears, for which they could discover no cause except my opinion.

"General Washington and his wife were both sanguine, except that she possessed a little of the encephalo-lymphatic. Napoleon Bonaparte was of the sanguine bilious encephalo-lymphatic constitution, and his first wife sanguine bilious lymphatic. General Jackson was sanguine bilious—the latter element predominating, and his wife purely bilious. In these instances, as it is well known, no children were produced. These historically known facts should have due weight in sustaining my observations, inasmuch as no explanation has hitherto been had of them. Of facts having this character, I have many which have passed under my own observation; two or three of which I will cite:

"Judge —— and his wife possessed the bilious-encephalic constitution; they lived to be old, but had no children.

"An intimate friend of the writer, a highly intellectual and educated gentleman, and his wife, one among the most polished of ladies, were of the sanguine encephalo-bilious constitution. They had no children --became at length mutually unhappy, and worse—mutually intemperate, and finally separated.

"An acquaintance of the writer married a number of years since, but had no children; his wife died, and three years since he married again. I told his brother, at the time, that he would have no children by his wife, or if he had he would not raise them. He asked me why, and I told him they were both alike— of the same constitution; when he remarked, 'That was the case with his first wife.' So far, the opinion I gave was correct.

"I know one party who were of the same constitution—had no children, and separated; but subsequently, each obtained a different companion and became blessed with children.

"These facts would seem to warrant some important conclusions with reference to the laws which govern in the premises, and to explain the fact that marriages in consanguinity degenerate the species; and yet, between the violation of the laws of temperate and consanguine marriages, there does not, to me, appear to be a general parallelism. The former does not appear to be attended with as much of what is usually called scrofula, or as much mental stupidity as the latter; but with more organic imperfection and liability to early disease and death. But upon this subject I am not confident; I want more facts; the evils attendant upon both may be referable to a violation of the same law under modifications of those conditions that constitute the premises.

"If we contemplate the two sexes as two distinct varieties of the human race, we shall find that one is but the complement of the other in all of the essential modes of existence. It does, then, absolutely require a good type of each to make one well-balanced and perfect unity. I have observed an exemplification of this law, in my intercourse in society, in those individuals who are so well developed and balanced in both body and brain as to be of no use to themselves or to others, unless impressed by external motives of such force and efficiency as to break up the prëexisting and normal balance of cerebral action. If, then, it be true that the good or best interests of the race require that one sex shall be the complement of the other—that the male shall have more developed than the female the animo-vital powers, and all those that act in concert with them—and the female shall have the vegeto-vital, and those powers that act in association with them, in a higher degree of endowment than the male—then it would seem very clearly to follow that the same law should be observed in all marriage alliances. When, there-fore, the animo-vital forces are strong in both parents, or feeble in both—or the vegeto-vital in a similar manner exist in both—then the law of the normal relations between the sexes has been violated; and in the ratio of the violation will be the serious-ness of the consequences.

"It may be inquired how the public are to be guided by the distinctions I have drawn, inasmuch as they are not understood by any one except those who have been my pupils? The answer is, that by these means they can not do it; but they can do that which is equivalent—they can determine how far they resemble, in complexion, make of person, and form of head, any other given individual, and how far they differ from him. If each man and woman will, as far as practicable, select their com-panions out of those who shall the least resemble themselves, they will very greatly but not entirely avoid the commission of a very serious blunder. The exceptions will be with those who, when young, are of a lean habit of body, but destined to acquire a full one; but even here the difficulty can be obviated by direct-ing attention to that side of the family of the individual which he or she may resemble.

"Among the forms of disease to which the violation of this law produces a liability is tubercular phthisis—one about which so much trouble has been had to trace each individual case to an hereditary origin. The success of this inquiry has not been found to extend beyond about thirty-six per cent., leaving sixty-four per cent. to be accounted for by some other means. As I have never known this form of disease to occur in any but those who are organically feeble in both vigor and tenacity of life, I concluded long since that a liability to it essentially inheres in such debility. Under the meteorological influences acting upon Boston, phthisis is a frequent disease; but those influences in the South produce, in the same class of organizations, intermittent fever, enlarged spleens, dropsy, and death. Hence, intermittent fever is just as hereditary in the South, as phthisis is in the North. The truth is, it is hereditary in no place; but a mere attendant upon certain forms of organization under prescribed circumstances. That such an organization is frequently hereditary—an heirloom in some families—I admit; but not more so than that which is liable to acute rheumatism, chronic rheumatism, and a host of other forms of disease. One thing is certain, phthisis occurs about twice as frequently without a consumptive parentage as with it; and in all such cases, I have no doubt, *through violations of the procreative or marriage laws.* If these views be sound, as I verily believe they are, the preventives are obvious—obedience to the laws of marriage and proper educational processes."

9

CHAPTER IX.

MARRIAGE AND LONGEVITY.

IT is generally conceded, by thinkers of both sexes, that married people live longer—as well as enjoy life more—than the unmarried. This settled conviction is the result of observation, unaided by the positive figures of statistics. But the latter has been supplied at last by the observations and comparisons, during a period of nine years, of the number of living population of Scotland with the number of deaths—made by Dr. James Stark, of Edinburgh, Corresponding Secretary and Principal Director of the General Register Office in that country.

The relative proportion between the death-rates of the married and of the unmarried is not absolutely uniform in different countries, but it is fairly represented by the following table, which exhibits the mortality per thousand of married and unmarried men in Scotland.

We regret that our limited researches have not enabled us to find any satisfactory tables or returns, of a similar character, about the United States, England or France.

Ages.	Husbands and widowers.	Unmarried.
20 to 25	6.26	12.31
25 to 30	8.23	14.94
30 to 35	8.65	15.94
35 to 40	11.67	16.02
40 to 45	14.07	18.35
45 to 50	17.04	21.18
50 to 55	19.54	26.34
55 to 60	26.14	28.54
60 to 65	35.63	44.54
65 to 70	52.93	60.21
70 to 75	81.56	102.71
75 to 80	117.85	143.94
80 to 85	183.88	195.40

From this table we are to understand that out of one hundred thousand living persons, of both sexes, it was found that in each quinquennial period, 597 married and 1174 unmarried died yearly between the ages of twenty and twenty-five ; between thirty and thirty-five, 865 married and 1396 unmarried died. The difference against unmarried life continued diminishing with the advance of age, until between the ages of seventy-five and eighty the deaths averaged yearly 1108 married, to 1454 unmarried. Among males older than twenty years, the average duration of life was 59.7 years for the married, and forty years for the single. Among those above twenty-five years old, the average was 60.2 years old for the married, and 47.7 for the single. Among females, the difference between fifteen and thirty, and between forty and forty-five years, slightly favored single life; from forty-five to ninety-five, it strongly favored marriage; and upon the whole time of life, marriage appeared to add to the longevity of women.

"The explanation of the facts indicated in these statistics, is simply that the additional labor laid upon a man by the burden of a family, is more than balanced by the restraint marriage sets up against vicious indulgences, which are prejudicial to longevity."

Or, to state it differently, Dr. Stark finds that the average age of married men over twenty years is over fifty-nine years, while the unmarried average only forty years; that is, marriage adds nearly one-third to the length of life, as a general rule, because, as another writer says—

1st. Bachelors are constantly in a State of unrest, they feel unsettled.

2d. If indoors after supper, there is a sense of solitariness, inducing a sadness, if not actual melancholy, with all their depressing influences; and many hours in the course of the year are spent in gloomy inactivity, which is adverse to a good digestion and a vigorous and healthful circulation.

3d. His own chamber or house being so uninviting, the bachelor is inclined to seek diversion outside, in suppers with friends, in clubs which are introductories to intemperance and licentiousness, or to those more unblushing associations which, under the cover of darkness, lead to speedy ruin of health and morals; and when these are gone, the way downward to an untimely grave is rapid and certain.

On the other hand, marriage lengthens a man's life—

1st. By its making home inviting.

2d. By the softening influences it has upon the character and the affections.

3d. By the cultivation of all the better feelings of our nature, and in that proportion saving from vice and crime.

4th. There can be no healthful development of the physical functions of our nature without marriage— it is necessary to the perfect man ; for Divinity has announced that it was "not good for man to be alone."

5th. Marriage gives a laudable and happifying object in life, the provision for wife and children, their present comfort and future welfare, the enjoyment in witnessing their happiness, and the daily and hourly participations in affectionate interchange of thought and sentiment and sympathy.

THE AGE AT WHICH PEOPLE MARRY.

The "age of consent" to marriage, by the common law, is fourteen for males and twelve for females. If a person of adult age marry one who is under the age of consent, such marriage is binding upon neither party ; and it is, by the rules of the common law, in the power of either to disagree when the infant comes to the age of consent, though not before.

Some of the United States, by their statute laws, fix the age of consent different from the common law. Ohio, for instance, declares marriages between males under the age of eighteen, and females under fourteen, invalid, unless confirmed by cohabitation after arriving at those ages respectively. Kentucky, however, only prohibits and declares void marriages between males of fourteen and females of twelve years of age, thus following and adopting the common law.

Among the Jews, in New Testament times, it was accounted scarcely reputable for a man to remain unmarried after eighteen years of age, and marriages in general were very early. In our day, in Electoral Hesse, Germany, marriage is prohibited to males under twenty-two, and to females under eighteen. A very natural result of this law, which in some cases seemed hard in its strictness, was that the girls frequently claimed to be eighteen long before attaining that age—the reverse of those in this country who, after passing their twenty-fifth or even their thirtieth year, claim still to be in their teens. The Hessian girls are in favor of early marriages.

In Wurtemburg, it is only by special dispensation that men are ever allowed to marry under the age of twenty-five. The law of Prussia makes the minimum age for marriages eighteen for a man, and fourteen for a woman. In Saxony, a man may not marry under twenty-one, while there is no restriction as to the age of a woman. A boy of fourteen may marry a girl of twelve in Austria, where the common law ages prevail. In Baden, twenty-five and eighteen are the respective ages, unless permission to marry younger is obtained from the police authorities. An ancient and curious law existing in Wurtemburg—which would prevent quite a large number of marriages in the United States if enforced here—forbids any woman to marry a man twelve years younger than herself.

Among the ancient Germans, than whom a finer race never existed, it was death for any woman to marry before she was twenty years old. By the laws of Lycurgus, the most special attention was paid to the physical education, and no delicate or sickly women were, on any account, allowed to marry. Dr. Johnston, in his work of " *Economy of Health,*" says that matrimony should not be contracted until the first year of the fourth septennial on the part of the lady, nor before the last year of the same in the case of the gentleman. In other words, the female should be at least twenty-one years of age, and the male twenty-eight years. The Doctor says there should be a difference of seven years between the sexes, at whatever period of life the connection is contracted. There is a difference of seven years, not in the actual duration of life in the two sexes, but in the stamina of the constitution, the symmetry of the form, and lineaments of the face. In respect to early marriage, so far as it concerns the softer sex, for every year at which marriage is entered upon before

the age of twenty-one, there will be, on an average,
three years of premature decay, more or less apparent, of the corporal fabric.

STATISTICS IN THE UNITED STATES.

But more interesting and curious is it to know,
not the ages at which men and women *may* by law
intermarry, but the ages at which they actually *do*
marry. In the city of Philadelphia, in 1861, 21.34
per cent. of brides were under twenty, while only 16
in all, or 0.36 per cent. of grooms were under twenty-
one years of age. 41.04 per cent. of brides, and
36.47 of grooms, were between twenty and twenty-
five years. 17.67 of brides and 30.17 per cent. of
grooms were between twenty-five and thirty years of
age. Of all the brides during that year, four-fifths
were under thirty, and two-thirds under twenty-five
years. In the same city, another year, while only
nineteen men were married under twenty years of age,
816 women were married who had not reached that
age. Nine men who were married were between
seventy and eighty years of age, and four women
were between sixty and seventy—thus seemingly
reversing the world-known aphorism about death,
" the young *may* die, but the old *must*," which, as to
marriage in Philadelphia, should read, " the old *may*
marry, but the young *do*."

In the State of Michigan, in the nine months
from April 5th to December 31, 1868, the number

of marriages was 5,485, or 10,970 persons married.
The following analysis shows the number and per
centage of persons married at certain periods of life:

	Males.	Females.	Percentage of Males.	Percentage of Females.
At 13 years.		2.		.03
" 14 "		8.		.14
" 15 "		58.		1.05
" 16 "		331.		6.03
" 17 "	5	398.	.09	7.25
" 18 "	37	631.	.67	11.50
" 19 "	105	589.	1.91	10.73
" 20 "	217	538.	3.95	9.80
Total under 20 yrs.	149	2020.	2.72	36.99
From 20 to 25 yrs.	2189	1963.	39.90	35.78
" 25 " 30 "	1605	723.	29.26	13.18
" 30 " 35 "	596	261.	10.86	4.75
" 35 " 40 "	314	163.	5.72	2.98
" 40 " 45 "	182	90.	3.31	1.64
" 45 " 50 "	126	73.	2.29	1.33
" 50 " 55 "	78	29.	1.42	.52
" 55 " 60 "	67	29.	1.22	.52
" 60 " 65 "	45	10.	.82	.18
" 65 " 70 "	25	7.	.45	.12
" 70 " 75 "	8	2.	.14	.03
" 75 " 80 "	6	1.	.10	.01
Over 80 years.	2		.03	
Unknown.	93	114.		
Total.	5,485	5,485		

The Secretary of State, in making his annual report, from which we have condensed the above, adds
another analysis, which we also condense, of thirty-
nine marriages—"exhibiting disparity of ages in

some parties, youthful ardor in many, mature consideration in others, and a high estimate of matrimonial alliances by all : "

Man aged 80, woman 30.	Man aged 46, girl 18.
" 80, " 71.	" 46, " 17.
" 79, " 52.	" 45, " 18.
" 78, " 66.	" 38, woman 60.
" 76, " 62.	" 31, girl 14.
" 75, " 67.	" 27, " 15.
" 75, " 58.	" 26, " 15.
" 75, " 50.	" 22, " 15.
" 74, " 75.	" 22, " 13.
" 74, " 68.	" 21, " 15.
" 74, " 53.	Boy aged 19, " 15.
" . 74, " 23.	" 19, " 13.
" 73, " 49.	" 18, woman 37.
" 73, " 50.	" 18, girl 16.
" 68, " 25.	" 18, " 16.
" 63, " 35.	" 18, " 16.
" 61, " 26.	" 18, " 15.
" 55, " 20.	" 17, " 17.
" 50, " 20.	" 17, " 16.
" 50, " 20.	

Thus, in Michigan, 6.63 per cent. of the males, or about 1 in 15 were married under the age of twenty-one years, while of females 46.58 per cent., or nearly 1 in 2$\frac{1}{5}$ were married before the age of twenty-one.

In Massachusetts, in 1867, but 1.6 per cent. of the males, and 19.9 per cent. of the females were married before the age of 21 years, exhibiting a very great disparity in the two states, and which is almost as marked in comparing all the old northern or north-eastern states with the new or border states and territories.

In the city of Boston, during the year 1862, there were 2,094 couples married. Of the grooms thirty-nine were under twenty-one; 729 (about 34.81 per cent.) between twenty-five and thirty years; 687 over thirty years—of whom two, each aged seventy-five, married women aged respectively twenty-five and seventy-eight years. Of the brides, 908 (43.40 per cent.) married between the ages of twenty-five and thirty; 337 (16.09 per cent.) over twenty-five years; fifty brides had to get the written consent of their parents because under age. 257 brides (12.27 per cent.) were older than their husbands; one man under thirty married a woman of fifty; four brides were only fifteen years old. Six *colored* men married white women. 82.14 per cent. of grooms, and 86.72 of brides were married for the first time; 325 men and 266 women were second marriages.

In the city of Cincinnati and Hamilton County, Ohio, during the year 1868, there were 2,797 couples married. Of these thirty-nine grooms were under twenty-one years of age, and seventy-two females under eighteen years of age.

STATISTICS IN ENGLAND.

In 1853, about 18 per cent. of the women married in England were under twenty-one years of age; in

1864 about 20 per cent., or one-fifth; and again in 1866 about 20 per cent.

From a record of several thousand cases of females under the care of Dr. Bennet, in the Western General Dispensary, London, we have ascertained the ages at which 237 of them were first married. From other sources we have obtained the data of 876 additional registered cases, making 1113 in all. This will enable the curious reader to ascertain the probable chances of marriage of English women at various ages. Thus, of these 1113 females, there were married:

3 at 13,	115 at 21,	27 at 28,	2 at 35,
13 at 14,	97 at 22,	23 at 29,	0 at 36,
19 at 15,	73 at 23,	7 at 30,	2 at 37,
51 at 16,	73 at 24,	7 at 31,	0 at 38,
69 at 17,	54 at 25,	7 at 32,	1 at 39,
89 at 18,	38 at 26,	8 at 33,	1 at 40,
140 at 19,	28 at 27,	6 at 34,	1 at 42.
145 at 20,			

CHAPTER X.

TWINS, TRIPLETS, AND EXTRAORDINARY FE-CUNDITY.

IT has been proven by microscopic observations that the semen of the male is filled with animalculæ shaped similar to the tadpole, each one of which, if healthy, and coming in contact with the healthy ovum of the female, is capable of generation. It is also claimed that at each menstrual period of the female, one or more eggs are expelled, and these, coming in contact with the animalculæ, are immediately impregnated by them. Leewenhock says " the seed of the male contains a multitude of spermatic animalculæ, all capable of becoming, by development, beings similar to their father. These animalcules push forward along the tubes, upon the ovaria, where a general engagement takes place, wherein all are slain but one, who, master of the field, finds the triumph of his victory within the ovum which has been prepared for him. According to this system, then, the survival of more than one of the combatants is the cause of twins." Whether the theory be well founded or not, we give it for what it is worth. It will amuse or interest the curious.

It not unfrequently happens that twins are born days or weeks apart. This may be caused at two separate copulations—the first just before or immediately after the menstrual flow, or at sometime within from eight to twenty days after the commencement of the flow. (It will be seen by reference to our table, pp. 84–5, showing the time conception takes place, that such a thing is not impossible in twenty-five days after the flow has commenced.)'

We append here a number of cases of *twins*, *triplets*, and *quaternions*; also, some remarkable cases of fecundation and generation.

TWINS.

Twins are of such frequent occurrence as, although always attracting special interest, seldom to occasion surprise. In some family connections, many households have their numbers thus increased; while in other extensive families, not a twin birth has ever happened.

In a very interesting record of 2,387 cases of consecutive labors, in the practice of Dr. H. Corson, of Montgomery County, Pennsylvania, as reported by Dr. C. to the Pennsylvania State Medical Society, out of 2,426 children in all, 78 were twins—about 3.21 per cent. 35 of these were males, 33 females.

In the state of Michigan, between April 5, 1868,

and December 31st, 1868, were born 210 pairs of twins—420 children—219 males, and 201 females. The rate per centum or proportion to the whole number of children born in the State during the same period, was 2.14—or 1 in 45⅔.

In the State of Vermont, in 1857, among 6,412 births were 92 twins—about 1.43 per cent. In 1858, of 6,477 births, 148 were twins—over 2.28 per cent. In 1859, of 6,545 births, 128 were twins —over 1.95 per cent.

Of the total number of births above mentioned, 40,992, there were 866 twins—about 2.11 per cent.

The Edinburgh *Medical Journal*, November, 1862, contains an account of a woman aged forty-seven, first pregnant at twenty-five, who has had fourteen pregnancies and twenty-five children—twins eleven times. Of these eleven cases of twins, in eight of them both children were born at the full time; in two cases, one fœtus was aborted about the third month, the other carried to the full time; and in one case, she miscarried both at the fourth month. Of the eight cases of twins born alive, the sex was boy and girl five times; both girls, twice; both boys, once. She had twins twice in eleven months and thirteen days. Her mother had twins once; but no other instance in the family.

Under the playful heading of " Never too late to do good," the New Albany (Indiana) *Ledger*, in 1867, recorded that the wife of an old citizen had just become the mother of two bouncing boys, weighing seventeen pounds, and added: "There is nothing particularly singular in a married lady becoming a mother. But in this instance there is something out of the 'usual course of things,' as the mother has reached the ripe age of fifty-seven years. Very few old maids could do as well, even if they had the opportunity."

Dr. Schroder describes a case in his own practice where twins were delivered six days apart, and both living.—*Braithwaite*, part v, p. 160.

The editor of the *British Record*, in his practice, experienced three cases where the births of twin children were separated by a space of time; in two cases an interval of twenty-one days each, and in the other fourteen days occurred. In all three cases, the first born fœtus was dead.—*Braithwaite*, part xviii, p. 284.

In the London *Lancet*, January, 1843, Mr. Vale recites the case of a woman, pregnant with twins, who gave birth to the first, alive, at the seventh month; and to the second, also alive, two months afterward, at the full period.

In Youngstown, Ohio, as recorded in the *Vindicator*, on June 30, 1870, a married lady gave birth to a child, remained in a feeble condition until the 6th of August, and then was delivered of a second child.

TRIPLETS.

Three sets of triplets were born in Michigan during the latter nine months of the year 1868—one boy and two girls in each of two cases, and two boys and one girl in the other. The mothers were born in Tennessee, Ireland, and Scotland, respectively. The proportion of triplets to the whole number of children born in Michigan was nearly 1 to 2,130.

Mr. Turton, in the *British Med. Journal*, May 2, 1868, mentions a case of a woman thirty-six years old, who, until then, had not been pregnant for fourteen years. After being ten hours in labor, she was delivered, at 2 A. M., February 18, 1868, of a living male child. About twelve hours after, when her husband came home to dinner, he found that, during the temporary absence of her attendants, she had just given birth to two lively girls. The physician was promptly called, and, other circumstances being favorable, she made an excellent recovery.

On February 11, 1868, Mrs. W., of Boone County, Indiana, aged thirty-five, was in labor with her second child, but added two more in less than twenty-five minutes—two boys and one girl—weighing six and one-half, seven, and eight pounds. A month after, all were doing well. Recorded in Cincinnati *Lancet*, September, 1868.

On July 11, 1867, Mrs. Sallie Royal, of Dooly County, Georgia, gave birth to three daughters, all living and doing well. She had been married two years, and had four daughters, not one of whom could walk alone.

A Philadelphia merchant informed the writer of a case within his personal knowledge, while a manufacturer at Nottingham, England. He loaned money to one of his factory hands, on condition he would, within three days, marry a young woman who had loved him, "not wisely, but too well." In one week after marriage, she was delivered of triplets, and, within eleven months afterward, of twins—all alive and well when our informant removed.

In July, 1868, the wife of Mr. Joseph Johnson, in Jefferson township, Morgan County, Indiana, gave birth to three children, and in less than eleven months thereafter, she gave birth to two more. The children were all perfectly formed; but only one of them survived a year afterward.

On the 4th of April, 1869, Mrs. John McDowell, of Robertson County, Kentucky, gave birth to three living male children, weighing twenty-three pounds. The largest weighed ten pounds.

On the 13th of May, 1869, the wife of Judge James Human, of Humansville, Polk County, Missouri, gave birth to two boys and a girl, averaging over six pounds each. The father of this trio of babies is sixty-nine years of age, has had three wives, and is the paternal relative of some twenty-five children.

The wife of a poor cab-driver, in London, in

the year 1868, gave birth to three children, one on the 5th, one on the 6th, and one on the 8th of May.

In January, 1869, Mrs. Cunningham, residing near Pigeon Run, in Campbell County, Virginia, gave birth to three fine boys, all alive and doing well. She was the second wife of her husband, he being seventy years of age, and she only about thirty-five.

In the year 1857, in the State of Vermont, only one set of triplets were born, all females, out of 6412 births, while the next year, out of 6477 births there were three sets of triplets, of which one set were females, still-born; another, all living, one male and two females; the other, all living on fourth day, one male, two females.

The *Am. Jour. Med. Sci.*, April, 1863, records a recent case of triplets in Liverpool, England, all living, strong children; two measured eighteen inches each in length, the third was one inch shorter.

The London *Medical Times*, April, 1862, records a case of seven children at three successive births— twins at each the first and second labors, then triplets—all born living, or speedily reanimated.

QUATERNIONS.

The circumstance of four children at one birth is quite rare, still there are many well-authenticated cases on record.

On the 7th of June, 1867, Mrs. Sarah Bush (col. ored), of Lexington, Missouri, was delivered of four healthy-looking children—the first two, girl and boy, weighing four and a half pounds each, and the last two, boys, weighing four and a quarter pounds each. Within a week after the girl and one boy died; the others were doing well.

In July, 1867, in Jasper, Alabama, the wife of William Hadnot (colored) presented him with four children—one weighing eight pounds, one twelve, one thirteen, and one fifteen—forty-eight pounds of babies at one birth! Even if William Had*not*, it was evident his wife had!

On April 1, 1867, Mrs. James Waters, living in Bonne Femme Bottom, Boone County, Missouri, was delivered, at a birth, of four boys, weighing six pounds each, " all alive and kicking, and it was n't a very good night for boys either." She had previously had six boys at three births, and the four just born made six boys *within less than one year!*

On August 21, 1868, Mrs. White, living ten miles from Spring Hill, Maury County, Tennessee (herself weighing two hundred and nine pounds, only a week previous), gave birth to four boys, well-formed and healthy, and weighing in the aggregate twenty-three pounds. She had been married about three years, and had a child fifteen months old.

Mrs. Joseph Murray, of Bluffton, Allen County, Ohio, in 1868, gave birth to twins, males, one still-born and weighing six and a half pounds, and the other weighing eight pounds. Just twenty-four hours afterward she gave birth to another pair of twins, males, weighing eight pounds each. The mother and three boys were thriving finely, a week later.

Dr. C. J. Faust, Oakland, Edgefield County, South Carolina, reports (*Am. Jour. Med. Sci.*, April, 1867, p. 564) the birth of four male children, one at 11 A. M., on February 26, 1867, and next morning, between 6 and 8 o'clock, three more, averaging five pounds, look well, seem perfectly healthy, and all nurse the mother. There was but one placenta, very large, square, and the umbilical cords were attached to each corner. The mother, just twenty-five years old, has had nine living children previously, and is doing as well as usual.

The London *Medical Times*, April, 1862, records a case of four children at a birth—two living, weighed five and six pounds, and two died, weighing three and a quarter and one and a half pounds. Once before, and once afterward, the same mother had twins.

FIVE CHILDREN AT A BIRTH.

Dr. Foote mentions the case of a woman of Lisle, who had five children at one birth. During the last two months of her pregnancy, according to the statement of the *Journal des Annonces*, all objects before her eyes were several times repeated, but after her delivery her sight returned to its natural state.

SIX CHILDREN AT A BIRTH.

There is a tradition in Italy that, in the sixteenth century, Dianora Frescobaldi had six children at one birth. If this be true, it is now too late to authenticate it, and it may as well remain *tradition !*

EXTRAORDINARY FECUNDITY.

The Wheeling (West Virginia) *Intelligencer* says that a gentleman of that city and a gentleman from New York, both of whom embarked upon the sea of matrimony about the same time, were talking of the prosperity of their respective families, when a third party, a mutual friend, remarked that he could tell a story of family prosperity which would beat them both. He then proceeded to state the following, which is certainly true: On July 24, 1858, a lady gave birth to a child; on May 30, 1859, she gave birth to two children; on May 29, 1860, she gave birth to two more; on May 1, 1861, she gave birth to three, and on February 15, 1862, she gave birth to four children, making twelve children in forty-two months and twenty days. The lady employs five nurses, whose duty it is to attend to these children. This story is vouched for by the best authority.

Mr. George Strodell, of Huntington, Indiana, when sixty-six years old, was the father of thirty-three children; and a local newspaper intimated that "coming events" continued to "cast their shadows before."

The Lancaster (Pa.) *Intelligencer*, of September 1, 1867, published the following: "On Thursday last, a German, named John Haeffler, living in Lancaster, Pennsylvania, followed to the grave his thirty-third child, and he was the father of thirty-seven children. Haeffler was married three times. His first marriage took place in Germany, when he was twenty-one years of age. With this wife he had seventeen children: at four successive accouchments triplets, twice twins, and the last a single birth. Shortly after the latter event the wife died. He was again married, and the issue of the second wife was fifteen children—seven times twins and the eighth a single child. This wife also died shortly after the last birth. His third and present wife has thus far presented him with five children, one at a birth. The sex of the children were nineteen boys and eighteen girls, only four of whom are now living—but whether the living are all issues of the last wife, or part of the previous wives, we have not ascertained. Haeffler is now fifty-two years of age, of medium size, and of hardy, vigorous constitution. In some sections of Germany a premium is awarded people who produce a certain number of children. Haeffler lived in that section, and was the recipient of one hundred guilders previous to leaving. He apparently labors under the impression that a similar reward awaits him in this country. If there is any State in the Union providing for such cases, the authorities had better send along the greenbacks.

In 1782, the name of Pheador Vacilitz was registered at Moscow. He was then seventy-five years old, and had been twice married. By his first wife he had sixty-nine children. She brought forth four children at a time in four births; three at a time in seven births; and twins on sixteen occasions. His second wife gave birth to eighteen children in eight deliveries, to twins in six, and to three children in two deliveries; so that in five and thirty years labor of his two wives, that peasant became the father of *eighty-seven* children, of whom eighty-three were living in 1782.

In the reign of the Empress Elizabeth, somewhat earlier in the last century than the instance above related, a peasant was brought to St. Petersburg and presented to Her Majesty. He was accompanied by eighty-two of his lawful children, and successfully applied for a pension.

Dianora Frescobaldi, an Italian lady of the sixteenth century, was the mother of fifty-two children. The inscription on her famous portrait by Bronzino in the San Donato collection, says that she never had less than three children at a birth, and there is a tradition in the Frescobaldi family that she once had six !

Brand, in his " History of Newcastle," mentions, as a well attested fact, that a weaver in Scotland had, by one wife, sixty-two children, all of whom lived to be baptized; and in Aberconway Church may still be seen a monument to the memory of Nicholas Hooker, who was himself a forty-first child, and the father of twenty-seven children by one wife.

In taking the New York State census, in 1865, fourteen American and nineteen foreign-born women were found who had borne twenty children each; and three American and two foreign women who had borne twenty-five children each.

TO PROPHESY THE NUMBER OF CHILDREN.

An old rule among the grandmothers of Southern Ohio—whence transmitted to them we have never heard—for ascertaining the number of children the future has in store for each family, is to count the wrinkles on the forehead of the father or mother; reckoning only the full wrinkles, or those which are well defined and reach across the forehead. We have heard of some quite amusing efforts at testing its reliability, at social assemblings of neighbors of both sexes, young and old.

CHAPTER XI.

ABORTION—STILL-BIRTHS—INFANTICIDE.

THE first two of these, in years gone by, were oftenest and properly known as misfortunes, seldom as crimes. In the demoralization which seems to characterize the last third of the nineteenth century—in all countries and among all people recognized as civilized or enlightened—they are oftener crimes than otherwise. We would not, for an instant, make the charge in stronger terms than the experience of all medical men and the actual sufferings of conscientious mothers fully justify. But there is very strong ground for believing that the larger portion of abortions, if not of still-births, in the United States are the result of intention, and not of accident.

Among physicians, it seems now to be the well settled opinion, that the foetus in utero is *alive* from the first moment of conception; and hence, that intentional dislodgment thereof, by whatsoever means produced, is highly criminal, if not positively murder. This dislodgment or premature expulsion of the fruit of conception every woman knows by the common term of abortion or miscarriage. Dr. H. R. Storer, of Boston, one of the ablest physicians of the age, in the Prize Essay, to which the American Medical Association awarded the gold medal for 1865, has so well and truly written upon this subject, that we think we serve the cause of humanity to present his views, on this point, in his own language and at length :

"Many women suppose that the child is not alive till quickening has occurred, others that it is practically dead till it has breathed. As well one of these suppositions as the other; they are both of them erroneous.

"Many women never quicken at all, though their children are born living; others quicken earlier or later than the usual standard of time ; or, others again may, in their own persons, have noticed either or all of these peculiarities in different pregnancies. Quickening is in fact but a sensation, the perception of the first throes of life—but of a twofold occurrence, and this not merely the motion of the child, but often the sudden emergence of the womb upward from its confinement in the low regions of the pelvis into the freer space of the abdomen. The motions of the child, which have been proved by Simpson, of Edinburgh, to be its involuntary efforts, through the reflex action of its nervous system, to retain itself in certain attitudes and positions essential to its security, its sustenance, and its proper development, are usually present for a period long prior to the possibility of their being perceived by the parent. They may very constantly be recognized by the physician in cases where no sensation is felt by the mother ; and the fœtus has been seen to move when born, during miscarriage, at a very early period.

"During the early months of pregnancy, while the fœtus is very small in proportion to the size of the cavity which contains it, sounds, produced by its movements, may be distinguished by the attentive ear applied to the abdomen of the mother, as gentle taps repeated at intervals, and continued uninterruptedly for a considerable time. These sounds may sometimes be heard several weeks before the usual period of the mother's becoming conscious of the motion of the child, and also earlier than the pulsations of the fœtal heart or the uterine souffle,[*] as the murmur of the circulation in the walls of that organ, or in the tissue of the after-birth, is technically termed. These motions must be allowed to prove life, and independent life. In what does this life really differ from that of the child five minutes in the world? Is not, then, forced abortion a crime? Moreover, instances have occurred where the membranes having been accidentally ruptured, the child has breathed, and even cried, though yet unborn—as proved alike by the sounds within the mother, well authenticated by bystanders, and by auscultation of her abdomen, and by the fact that sometimes, when not born living, the lungs of the fœtus have been found fully expanded, a process which can be effected only by respiration, and of which the proofs are such as can be occasioned in no other way whatever.

"In the majority of instances of forced abortion, the act is committed prior to the usual period of quickening. There are other women, who have confessed to me that they have destroyed their children long after they have felt them leap within their womb. There are others still, whom I have known to willfully suffocate them during birth, or to prevent the air from reaching them under the bedclothes; and there are others, who have willfully killed their wholly separated and breathing offspring, by strangling them or drowning them, or throwing them into a noisome vault. Wherein among all these criminals does there in reality exist any difference in guilt?

" That there has existed a wide and sincere ignorance of the true character of the act, I have allowed. At present let us turn from the crime against the child, to the crime as against the mother's own life and health: I refer more particularly to her own agency therein. Of the guilt of abortion when committed by another person than herself, and with reference both

---

[*] Naegele: *Treatise on Obstetric Auscultation*, p. 50.

to the mother's life and that of the child, there can be no doubt; but it is to the woman's own agency in the act, as principal, or accessory by its solicitation or permission, that we have now to deal; not as to its abstract wrong alone, but as to its physical dangers, and therefore its utter folly.

"It is generally supposed, not merely that a woman can willfully throw off the product of conception without guilt or moral harm, but that she can do it with positive or comparative impunity as regards her own health. This is a very grievous and most fatal error, and I do not hesitate to assert, from extended observation, that, despite apparent and isolated instances to the contrary—

"1. A larger proportion of women die during or in consequence of an abortion, than during or in consequence of childbed at the full term of pregnancy.

"2. A very much larger proportion of women become confirmed invalids, perhaps for life; and,

"3. The tendency to serious and often fatal organic disease, as cancer, is rendered much greater at the so-called turn of life, which has very generally, and not without good reason, been considered as especially the critical period of a woman's existence.

"These are conclusions that can not be gainsaid, as they are based on facts; and that these facts are merely what ought, in the very nature of things, to occur, can readily enough be shown.

"1. Nature does all her work, of whatever character it may be, in accordance with certain simple and general laws, any infringement of which must necessarily cause derangement, disaster, or ruin. It has been ascertained, by careful dissections and microscopic study, that the woman's general system, both as a whole and as regards each individual organ and its tissues, is slowly and gradually prepared for the great change which naturally occurs at the end of nine months' gestation; and that if this change is by any means prematurely induced, whether by accident or design, it finds the system unprepared. Not even do I except from this law the earlier months of pregnancy, when it is thought by so many that abortion can be brought on without any physical shock. During pregnancy, all the vital energies of the mother are devoted to a single end: the protection and nourishment of the child. Such wise provision is made for its security, such intimate vascular connection is established between the fœtal circulation and the blood-vessels of the mother, that its premature rupture is usually attended by profuse hemorrhage, often fatal, often persistent to a greater or less degree for many months after the act has been completed, and always attended with more or less shock to the maternal system, even though the full effect of this is not noticed for years.

"In birth at the full period, it is found that what is called by pathologists fatty degeneration of the tissues, occurs both in the walls of the mother's womb, and in the placenta or afterbirth, by which attachment is kept up with the child. This change, in all other instances a diseased process, is here an essential and healthy one. By it the occurrence of labor at its normal period is to a certain extent determined; by it is provision made against an inordinate discharge of blood during the separation and escape of the after-birth, and by it is the return of the uterus to the comparatively insignificant size, that is natural to it when unimpregnated, insured. Any deviation from this process at the full term, which prevents the whole chain of events now enumerated from being completed, lays the foundation of, and causes a wide range of uterine accidents and disease, displacements of various kinds, falling of the womb downwards or forwards or backwards, with the long list of neuralgic pains in the back, groins, thighs, and elsewhere that they occasion; constant and inordinate leucorrhœa; sympathetic attacks of ovarian irritation, running even into dropsy, etc. These are only a portion of the results that might be enumerated.

"Now, while all this is true of any interference with the natural process at the full time, it is just as true, and, if any thing, more certain, when pregnancy has been prematurely terminated and out of many hundred invalid women, whose cases I have critically examined. in a very large proportion I have traced these symptoms directly back to an induced abortion.

"Again: not merely does nature prepare the appendages of the child and the womb of its mother for the separation that in due time is to ensue between them, it also provides an additional means of insuring its successful accomplishment through the action that takes place in the woman's breasts, namely, the secretion of the milk. Though the escape of this fluid does not ordinarily occur in any quantity until some little time after birth has been effected, yet the changes that ensue have gradually been progressing for days, or weeks, or even months; for, as is well known, in some women the lacteal secretion is present before birth, at times even during a large part of pregnancy; and in all women there is, doubtless, a decided tendency of the circulation toward the breasts prior to the birth of the child—just as there has been so extreme a tendency of the circulation for so long a time toward the womb. It is, indeed, to take the place of the latter that the former is established, and to prevent the evil consequences that might otherwise ensue. The sympathy between the mammary glands and the uterus is now well established; it is shown in many different ways: in some women the application of the child to the breasts is immediately followed by after-pains, and in others, these pains, which are usually but contractions of the womb, to expel any clots that may have accumulated, are attended by a freer secretion or discharge of the milk. It is not uncommon, when the monthly discharge is scanty or suddenly checked, for the breasts to become enlarged and painful, as is so often the case soon after impregnation; while, on the other hand, one of the most efficient means we have of establishing the periodical flow, when suppressed, is by the application of sinapisms to the surface of the breasts. In view of these facts, it will readily be understood why it is that women who make good nurses are so much less likely than others to suffer from the various disorders of the womb, and why they are also less likely to rapidly conceive, and why, moreover, too long lactation should not be indulged in for either of these so desirable ends. The demands of fashion shorten or prevent nursing; the demands of fashion often forbid a woman from bearing children; but whether this is attained by the prevention of impregnation, or by the induction of miscarriage, it is almost inevitably attended—as is, to a certain extent, the sudden cessation of suckling—by a grievous shock to the mother's system, that sooner or later undermines her health, if even it does not directly induce her death.

"Dangers attend the occurrence of abortion which directly threaten a mother's life. This is true of all miscarriages, whether accidental or otherwise; but these dangers are enhanced when the act is *intentional*. When caused by an accident, the disturbance is often of a secondary character, the vitality of the ovum being destroyed, or the activity of the maternal circulation checked, before the separation of the two beings from each other finally takes place. But in a forced abortion there is no such preservative action; the separation is immediate if produced by instruments, which often besides do grievous damage to the tissues of the mother with which they are brought into contact, lacerating them, and often inducing subsequent sloughing or mortification; or, if the act is effected by medicines, it is usually in consequence of violent purgation or vomiting, which, of themselves, often occasion local inflammation of the stomach or intestines, and death. Add to this that even though the occurrence of any such feeling may be denied, there is, probably, always a certain measure of compunction for the deed in the woman's heart—a touch of pity for the little being about to be sacrificed; a trace of regret for the child that, if born, would have proved so dear; a trace of shame at casting from her the pledge of a husband's or lover's affection; a trace of remorse for what she knows to be a wrong, no matter to what small extent, or how justifiable it may seem to herself—and we have an explanation of the additional element in these intentional abortions, which increases the evil effect upon the mother, not as regards her bodily health alone, but, in some sad cases, to the extent even of utterly overthrowing her reason.

"The causes of an immediately or secondarily fatal result of labor at the full period are few; in abortion nearly every one of these is present, with the addition of others peculiar to the sudden and untimely interruption of a natural process, and the death of the product of conception. There is the same or greater physical shock, the same or greater liability to hemorrhage, the same and much greater liability to subsequent uterine or ovarian disease. To these elements we must add another, and by no means an unimportant one: a degree of mental disturbance, often profound, from disappointment or fear, that to the same extent may be said rarely to exist in labors at the full period.*

"Viewing this subject in a medical light, we find that death, however frequent, is by no means the most common or the worst result of the attempts at criminal abortion. This statement applies not to the mother alone, but, in a degree, to the child. Many of the measures resorted to are by no means certain of success, often, indeed, decidedly inefficacious in causing the immediate expulsion of the fœtus from the womb; though almost always producing more or less severe local or general injury to the mother, and often, directly or by sympathy, to the child.

" The membranes or placenta may be but partially detached, and the ovum may be retained. This does not necessarily occasion degeneration, as into a mole, or hydatids, or entire arrest of development. The latter may be partial, as under many forms, from some cause or another, does constantly occur; if from an unsuccessful attempt at abortion, would this be confessed, or, indeed, always suggest itself to the mother's own mind? Fractures of the fœtal limbs, prior to birth, are often reported, unattributable in any way to the funis, which may amputate, indeed, but seldom break a limb. A fall or a blow is recollected; perhaps it was accidental, perhaps not, for resort to these for criminal purposes is very common. In precisely the same manner may injury be occasioned to the nervous system of the fœtus, as in a hydrocephalic case long under the writer's own observation, where the cause and effect were plainly evident. Intra-uterine convulsions have been reported; as induced by external violence, they are probably not uncommon, and the disease, thus begun, may eventuate in epilepsy, paralysis, or idiocy.

* Studies of Abortion : *Boston Medical and Surgical Journal*, February 5, 1863.

"To the mother there may happen correspondingly-frequent and serious results. Not alone death, immediate or subsequent, may occur from metritis, hemorrhage, peritonitic, or phlebitic inflammation, from almost every cause possibly attending, not merely labor at the full period, comparatively safe, but mis-carriage increased and multiplied by ignorance, by wounds, and violence; but if life still remain, it is too often rendered worse than death.

"The results of abortion from natural causes, as obstetric disease, separate or in common, of mother, fœtus, or mem-branes, or from a morbid habit consequent on its repetition, are much worse than those following the average of labors at the full period. If the abortion be from accident, from ex-ternal violence, mental shock, great constitutional disturbance from disease or poison, or even necessarily induced by the skillful physician in early pregnancy, the risks are worse. But if, taking into account the patient's constitution, her pre-vious health, and the period of gestation, the abortion has been criminal—these risks are infinitely increased. Those who es-cape them are few.

"In thirty-four cases of criminal abortion reported by Tar-dieu, where the history was known, twenty-two were followed, as a consequence, by death, and only twelve were not. In fifteen cases necessarily induced by physicians, not one was fatal.

"It is a mistake to suppose, with Devergie, that death must be immediate, and owing only to the causes just mentioned. The rapidity of death, even where directly the consequence, greatly varies; though generally taking place almost at once if there be hemorrhage, it may be delayed even for hours where there has been great laceration of the uterus, its sur-rounding tissues, and even of the intestines; if metro-peritonitis ensue, the patient may survive for from one to four days, even, in-deed, to seven and ten. But there are other fatal cases, where on autopsy there is revealed no appreciable lesion, death, the pen-alty of unwarrantably interfering with nature, being occasioned by syncope, by excess of pain, or by moral shock from the thought of the crime.

"That abortions, even when criminally induced, may sometimes be safely borne by the system, is of little avail to disprove the evidence of numberless cases to the contrary. We have instanced death. Pelvic cellulitis, on the other hand, fistulæ, vesical, uterine, or between the organs alluded to; adhesions of the os or vagina, rendering liable subsequent rupture of the womb during labor or from retained menses, or, in the latter case, discharge of the secretion through a Fallopian tube, and consequent peritonitis; diseases and degenerations, inflammatory or malignant, of both uterus and ovary; of this long and fearful list, each, too frequently incurable, may be the direct and evident consequence, to one patient or another, of an intentional and unjustifiable abortion.

"We have seen that, in some instances, the thought of the crime, coming upon the mind at a time when the physical system is weak and prostrated, is sufficient to occasion death. The same tremendous idea, so laden with the consciousness of guilt against God, humanity, and even mere natural instinct, is undoubtedly able, where not affecting life, to produce insanity. This it may do either by its first and sudden occurrence to the mind, or, subsequently, by those long and unavailing regrets, that remorse, if conscience exist, is sure to bring. Were we wrong in considering death the preferable alternative?*

"To the above remarks it might truthfully be added, that not only is the fœtus endangered by the attempt at abortion, and the mother's health, but that the stamp of disease thus impressed is very apt to be perceived upon any children she may subsequently bear. Not only do women become sterile in consequence of a miscarriage, and then, longing for offspring, find themselves permanently incapacitated for conception, but, in other cases, impregnation, or rather the attachment of the ovum to the uterus, being but imperfectly effected, or the mother's system being so insidiously undermined, the children that are subsequently brought forth are unhealthy, deformed, or diseased. This matter of conception and gestation, after a miscar-

---

* *Criminal Abortion in America*, p. 42.

riage, has of late been made the subject of special study, and
there is little doubt that from this, as the primal origin, arises
much of the nervous, mental, and organic derangement and
deficiency that, occurring in children, cuts short or embitters
their lives.

"It may be alleged by those who, skeptical or not skeptical
as to these conclusions, have reason, nevertheless, to desire
to throw discredit upon them, that the weekly or annual
bills of mortality, the mortuary statistics, do not show such
direct influence from the crime of abortion as I have claimed
exists.

"On the other hand, it must not be forgotten that in these
cases there is always present every reason for concealment. In
the earlier months of pregnancy it is very difficult to prove, in
the living subject, that pregnancy has occurred. Such a conclusion being arrived at, before the sound of the fœtal heart can
be heard, for this is the only sign that is positively certain, by
merely circumstantial and probable evidence, which becomes of
weight only as it is accumulated and found corroborative. In
the dead subject, the victim of an abortion in the earlier
months, the case is often equally obscure, or at least doubtful,
unless the product of conception has not yet escaped, or, having been thrown off, has been detected or preserved. When
found, it of course proves pregnancy, whether the parent be
living or dead; that is, in the former instance, if its discharge
can be traced directly to the woman in question, and to no
other, and correlative circumstances may show that an abortion
has occurred; but this may have been accidental and guiltless.
Where the act has been committed by an accomplice, the proofs
of such commission and of the intent, though this is generally
implied by the act itself, are by no means always forthcoming.
Where the abortion has been induced by the woman herself, as
is now so frequently the case, certainty upon the point becomes
far more difficult. The only positive evidence by which to
judge of the real frequency of the crime is *confession*, and it is
from the confessions of many hundreds of women, in all classes
of society, married and unmarried, rich and poor, otherwise

good, bad, or indifferent, that physicians have obtained their knowledge of the true frequency of the crime.

"The confidential relations in which the physician stands to his patient; the understanding that nothing can wring from him her disclosures, save the direct commands of the law, so unlikely in any given case to become cognizant of its existence, elicits from a woman, in almost every instance, especially if she believes herself in peril of death, a frank statement of the means by which she has been brought low; for it is evident that upon such knowledge must depend the measures of relief to which the physician may resort. Could the test of confession be always applied, as is, however, manifestly impossible, so many women die during or in consequence of an abortion, without the attendance of a physician and without making any sign, it would be found that many of the cases now reported upon our bills of mortality as deaths from hemorrhage, from menorrhagia, from dysentery, from peritonitis, from inflammation of the bowels or of the womb, from obscure tumor, or from uterine cancer, would be found in reality to be deaths from intentional abortion. At first sight it would seem impossible that such grossly erroneous opinions as the above could be rendered; but their likelihood is readily perceived when it is recollected how often, when the best medical skill has been secured, attending circumstances are such as to excite little or no suspicion of the true state of the case, and a physical examination of the patient is therefore neglected. Women are still allowed to die of ovarian or of other tumors that might be easily and successfully removed, and, in default of a proper examination, are sometimes mistakenly pronounced instances of disease of the liver or of ordinary abdominal dropsy, and as such are buried. If such and similar errors can occur in chronic cases, where time and opportunity have permitted the most thorough examination and study, still more likely are they to take place during the hurry and anxieties of an acute and alarming attack, where the conscience and shame of the patient are alike interested in causing or keeping up a deception.

"It will have been seen, then, not merely that an induced abortion may be attended with great immediate danger to the mother, but that in reality it is very often fatal, either from the so-called shock to her system, or from hemorrhage, or from immediately ensuing peritonitis.

" 2. Should the woman survive these immediate consequences, no matter how excellently she may have seemed to rally, she is by no means safe as to her subsequent health. There are a host of diseases, some of them very dangerous, to which she is directly liable.

"The product of conception is not always entirely gotten rid of. If a fragment remains, no matter how trifling in size, it may serve as the channel of the most severe and constant hemorrhagic discharge. Of this, examples are by no means infrequent; the flux lasting at times for very many months, and, if the cause is not finally detected and removed, hurrying the patient to her grave.

"The product of conception is sometimes retained entire, after its detachment from the uterine walls has been supposed wholly effected. It may be carried for many years, always acting as a foreign body; at times occasioning extreme irritation, shown perhaps only by distant and otherwise inexplicable symptoms, or it may lie dormant for a time without apparent trouble—finally making itself known by some sudden explosion of disease, whether by purulent absorption and general pyæmia; by ulceration and discharge of fœtal debris, through the intestines, bladder, or even abdominal integuments; or, by metritic inflammation, followed by sympathetic or consequent fatal peritonitis.

"The patient, after an abortion, is very liable to one or another of the forms of uterine displacement, which are now known to lie at the foundation of so very large a proportion of the lame backs, formerly supposed consequent on spinal irritation; of the painfully neuralgic breasts, so often suggestive of incipient cancer; of the disabled limbs, pronounced affected with sciatica, cramps, or even paralysis; of the impatient bladders, from whose irritability or incontinence the kidneys are supposed diseased; of the obscure abdominal aches and pains, which unjustly condemn so many a liver and so many an ovary; of the constipation from mere mechanical pressure, which is so often thought to argue stoppage from stricture or other organic disease; of the severe and intractable headaches that, resisting all and every form of direct or constitutional treatment, are supposed to indicate an incurable affection of the brain; of the easily deranged stomachs, that are so suggestive of ulceration or of malignant degeneration; of the general hypochondria and despondency, that of the most gentle, even almost angelic, dispositions make the shrew and virago, and of the purest and most innocent produce, in her own conceit, the worst of sinners, even at times effecting suicide. Who that has suffered will think this picture overdrawn? Who that has practiced will not recognize in displacements the key by which these riddles may be solved?

"Their mode of causation is plain. After an abortion, just as after labor at the full term, the womb is more weighty than natural—its walls thicker and heavier than usual, alike by the excess of blood they contain, and by the increased deposition of muscular fiber. After childbed it has been shown that this increase is normally lessened by certain physiological processes attending the natural completion of that function. After an abortion these processes are absent or are but imperfectly performed. It is notorious that during the slight increase of weight from simple congestion that occurs at the regular monthly periods, women are very liable to displacement on any effort, extreme or slight, whether riding on horseback, gently lifting, or even straining at stool; during or after an abortion the risk is very greatly increased.

"With equal justice could I refer to the chances of trouble that otherwise accompany the premature ending of pregnancy. In many instances, I have now been summoned to attend, and frequently to operate upon, the consequences of local uterine or vaginal inflammation or of laceration, for both of these results may ensue where the womb has not been prepared to evacuate itself by the normal closure of pregnancy—and this, whether

or not instruments may have been employed. Adhesions of
varying situation and extent are not uncommon, as the result of
an abortion. They may be slight, and merely tilt or draw the
womb to one side, giving rise only to severe local or distant
neuralgias, and rendering the occurrence of a subsequent pregnancy somewhat dangerous; they may be more decided, and as
bridles or septa partially close the canal of the vagina, rendering menstruation and conjugal intercourse alike difficult and
painful; they may be so complete as entirely to obliterate the
mouth of the womb or of the external passage, in these instances preventing the escape of the menses, and rendering an operation necessary to avoid a rupture that might perhaps be fatal.
Should it be the outer entrance that is occluded, the woman is,
of course, entirely shut off from her husband's embrace; an
effect that, however grateful to many an invalid, her shame
would hardly be willing to accept as the consequence of disease.

"These are but a tithe of the pathological effects daily revealed to physicians, as in consequence of an intentional abortion.

"3. But not only is a woman in peril both as to life and health,
alike at the time of an abortion and for months or years subsequently. She may seem to herself and to others successfully
to have escaped these dangers, and yet when she has reached
the critical turn of life, succumb.

"At this eventful period, when the fountains of youth dry
up, and the scanty circulation is turned from its accustomed
channel, the woman ceases from the periodical discharges,
which in health and with care are the secret of her beauty,
her attractions, her charms. At its occurrence not merely is a
change produced in the system generally, but the womb, no
longer required, becomes atrophied and dwindles into insigni-
ficance. It may have had impressed upon it, years and years
back, the stamp of derangement, till now not rendered effective;
for, as in other portions of the body, a part once weakened may
retain itself in tolerably good condition until some accident or
other change develops or awakens the seed of disease.

"Little the comfort for a woman to have had her own way against the dictates of her conscience, the advice, perhaps, of her physician, if to the dangers she must directly incur, she must add the looking forward through all the rest of her life to possible disease, invalidism or death as the direct consequence of her folly; no wonder if she should consider prevention better than such cure as this, and yet the prevention of pregnancy, by whatever means it may be sought, by cold vaginal injections or by incomplete or impeded sexual intercourse, is alike destructive to sensual enjoyment and to the woman's health; her only safeguard is either to restrict approach to a portion of the menstrual interval, or to refrain from it altogether.

"Not merely are certain of the measures to which I have alluded detrimental to the health of the woman; they are so to both parties engaged, and it is to their frequent employment, freely confessed as this is to the physician, that much of the ill health of the community, both of men and women, is to be attributed. Though they may seem sanctioned by the rites of marriage, they are in some respects worse for the physical health, I might almost say for the moral health likewise, than illicit intercourse or even prostitution, for they bring both parties down to all the evils and dangers, mental and physical, of self-abuse."

But as figures and statistics have an influence with thinking persons, as well as casual readers, far greater usually than the experience and observations of medical men, unsustained by figures, let us see what this favorite mode of teaching would spread before us:

In the city of New York, from the record of the coroner's inquests for the year preceding October 1, 1867, it is ascertained—so we gather from an industrious anonymous writer—that one hundred and forty-three inquests were held on infants and fœtuses which had been cast into the streets, lots, cemeteries, and parks of this city. Of these, sixty-four had passed embryotic state and been born alive, according to the best medical testimony. Eight were obviously strangled, six suffocated, and one poisoned. Even around the neck of one was the cord which had drawn out its life. The remainder died of the ordinary neglect and exposure, and the return was the chimerical rendering of death by " unknown causes." These comprise the number found dead, actually dead, in one year by only one branch of the service in the city government. " Found dead," of course, does not include the whole number. Dead, but not "found," will comprise many more. The undiscovered murdered ones, for concealment is a leading element in all such transactions, are probably more than what are brought to light.

But besides this large number of infants absolutely killed, either before birth, or during, or shortly after birth, there is another large class who are picked up in the streets, found alive somewhere, but nearly all of whom die in a few days or months. Of these there were in 1865 picked up 153 infants, 149 in 1866, and 176 in 1867—a total in three years of 478. Add this to the one hundred and forty-three found dead in the last named year(1867), and the reader has the irreversible and undeniable official fact that two hundred and forty living-born infants, destroyed or abandoned, fell into the hands of the police in one year. And then, to either the medical or the humane reader, the seventy who had not advanced beyond the embryotic, tell a more awful story.

Of those who are picked up, disposition is made by sending them to the Foundling Hospitals. In these, despite, perhaps, careful and even sympathetic care, what with the ravages of previous exposure and the absence of the mother's indispensable nourishment, *four out of every five die.*

Out of one thousand deliveries recorded by Dr. M. C. Richardson, of Hollowell, Maine, there were born 967 living children, and 45 still-born—a little over 4½ per cent. He adds: "Cases of abortion occur too frequently in this community, and, strange as it may seem, most of them are among the married, brought on in many instances to avoid having large families. These cases are, in general, attended with more flooding and greater prostration, and a much larger proportion of deaths, than follow confinement at the full term with nursing."

In 1,000 cases under the care of Dr. Lawrence, of Montrose, according to his report, 45 were still-born —just 4½ per cent.

In Massachusetts, in 1867, the returns show 1,007 still-births to 35,062 living births—2.79 per cent. As this is less than two-thirds the rate shown in the two accurate reports just preceding this, and less than the rate in other full reports, it is evident that the return of still-births in that State was incomplete.

An interesting article in the *American Medical Times* for 1863, p. 153, says the proportion of still-births to the living gives the only basis on which can be calculated the number of cases of abortion. These figures are only approximative—for very many cases of still-birth are not produced by abortions, while a vast number escape detection and registration. Taking our mortality reports, with all due allowances for these discrepancies, the record is still sufficiently humiliating. Since the first registry in New York in 1805, the proportionate and actual increase of still-births has been alarmingly rapid. In 1805, the ratio of fœtal deaths to the population was 1 to 1,633; but in 1849, 1 to 340. In 1856, the records show that 1 in every 11 is still-born in New York City, while reports of European countries, even allowing for criminal abortions, give the proportion of still-births at 1 in every 15. Accurate records of the best practitioners give as the ratio of premature births, or non-viable fœtuses, to the whole number of births (which includes, of course, only abortions from natural or accidental causes), 1 to 78 ; but in New York, the ratio of the same births to the whole number, is 1 to 40. The ratio of premature still-births at full time, in New York City, in 1846, was 1 in 10; and in 1856, ten years later, it had increased to 1 in 4. From these facts it is apparent—not only that produced abortions are frequent in that community, but that they are rapidly increasing. In seven years, from 1850 to 1857, the still-births doubled; and we have good evidence that since that time the proportion has rapidly increased.

The registration returns of the State of Massachusetts show that the comparative frequency of abortions in that State, is thirteen times as great as in New York City.

A patient statistician has carefully prepared the appended tables to exhibit the facts and the comparisons of this subject in New York and in Paris:

FOR NEW YORK.

Years.	Deaths.	Still-born.	Ratio.
1851	21,748	1,286	6.1
1852	20,296	1,405	6.9
1853	21,137	1,575	7.5
1854 (Cholera year.)	26,953	1,615	6.0
1855	21,478	1,564	7.3
1857	21,775	1,553	7.2
1859	21,645	1,331	6.2
1867	23,443	2,228	9.5

The first thing that strikes the mind is the enormous increase of still-born children in the year 1867 over any of the years from 1851 to 1859. The explanation can only be found in the fact that the crime of abortion has obtained a prominence which makes it a profession and an immunity, and, therefore, profitable and hardly condemned. Now look at Paris on the same subject, though not wholly during the same years:

Years.	Deaths.	Births.	Still-born.	Still-born to 100 births.	Still-born to 100 deaths.
1851	42,521	38,342	2,867	6.7	7.5
1852	44,538	36,740	3,032	6.8	8.3
1853	46,707	44,330	3,171	6.8	7.2
1854	50,708	55,244	3,433	6.8	6.2
1855	49,688	49,366	3,219	6.5	6.5
1856	54,520	41,985	3,782	6.9 ·	9.0

In interpreting the above tables it should be borne in mind—

That the ratio of still-born children to one hundred living births in Paris, is twice greater than in any other part of France, and far greater than even in Belgium, the most densely peopled country in Europe.

That the ratio of still-born births to living births is twice as great in France among illegitimate as among legitimate cases.

That more than one-fourth of the births which occur in Paris every year are *illegitimate.*

That in Paris the foundling and lying-in hospitals allure thousands of women from the neighboring departments because of the freedom from exposure they assure, and the provisions therein made for illegitimate children.

These considerations tell in favor of New York so far as comparison goes.

But, as if this story of shame and crime in New York were not enough, it is known to the police and to the Board of Health that there are 120 persons whose business is the destruction of infants, and who, in the means to accomplish that, incidentally and necessarily, in numberless cases, destroy maternal life, also. They are known as the sellers of French, Portuguese, and other "female pills," keepers of "confidential" medical offices, etc. They can not be convicted of the crimes alluded to, except upon proof; and the very subjects of their crimes—because accessories in the eye of the law, and subject to punishment—are safely relied upon to keep the villainous secret from the officers of the law and from all who would expose them. What becomes of the fœtuses thus ruthlessly dragged at once out of existence and into the world, has not positively been ascertained, although reasonable conjecture adds to the crime itself the more horrible barbarity that the often unshapely bodies have been *burned* in these dens of death and sepulchers of shame—seemingly upon a principle more painfully true than the murderer's comfort, that "dead men tell no tales." If the jurisdiction of the self-designated officers of Judge Lynch's court and law, with its trials and executions alike without a hearing or time for repentance, could only be extended for a few days to this class of murderers in New York and elsewhere, there would be such a quaking of dry bones as few good people would ever regret or inquire into.

The head of the bureau of Vital Statistics, of the New York Board of Health, Dr. Harris, with all the revealed facts and the outgivings of more painful mysteries before him, has recorded his opinion

that the actual number of still-born children and
of those picked up dead is two-thirds greater than
the published reports show. If, instead of his
large estimate, it be only twice as great, it follows
that more than 2,000 mothers in our great metrop-
olis every year destroy or procure others to destroy
their own offspring. What a terrible commentary
this upon the morals of those who are guilty of,
or aid and abet, this unnatural aversion of women to
become mothers. The husbands, seducers, medical
advisers, sympathizers who assist in and in covering up the crime of fœticide, would number 10,000
at least; while in ten years, at the ratio given, even
supposing many females repeat the dangerous experiments upon their own and their children's lives,
at least 18,000 women are guilty of that crime
against themselves, society, and God.

A published report, in 1868, made by the officers
of the Sanitary Association of Montreal, Canada,
of the operations of the *Foundling Hospital* of that
city, under the management of the *Gray Sisters*,
gives the whole number of children received at the
hospital during the year as 652, of whom 413
were born in Montreal, and the remainder in other
places, 29 being sent from the United States. The
accounts of the treatment of some of those from the
United States are revolting. The report says "one
was sent in a carpet bag, another in a basket, one at
the bottom of a water bucket, one strongly nailed up
in a box, two squeezed and bruised, and another

with a pin stuck through its flesh." 424 infants were sent to the hospital only half clothed, and 8 came entirely naked into the hands of the Sisters. 18 had not even been washed; 13 were bleeding from want of proper treatment at birth; 46 were tainted with an infamous disease, derived from their mothers; 8 had been injured by instruments; 7 were more or less frozen, and large numbers were covered with vermin; 3 were dead when received; 23 were dying, and 157 in actual disease. The deaths during the year were, of course, frightful, amounting to 619, leaving a balance of only 33 to reward the humane exertions of the Sisters. Of the dead, 36 were under one week old; 368 under one month; 583 under one year; 617 under five years of age.

The mortality among infants is terrible. It is well established that, under ordinary contingencies, about $3\frac{1}{2}$ per cent. of all births are dead-born; but statistics teach undeniably that of recent years asserted cases of this kind are far more numerous, averaging about 10 per cent. In addition to these, a large number of children die within a few weeks or months after birth. In Philadelphia, in the year 1867, five-sixths per cent. of the whole number of births were still-born. Of those born alive in that city, the statistics of the Board of Health show that 27 per cent. died before attaining one year, and 11 per cent. of the remainder in the fol-

12

lowing year. At the age of 20, half of those born alive would be dead.

From careful statistics presented by the Springfield *Republican,* it would seem that Massachusetts gets rid of an extraordinary number of children that are permitted to come to birth. There are no means of knowing precisely how many ante-natal murders are committed in that State every year, and an estimate only can be made by showing that the increase in population is due almost wholly to the foreign-born and foreign-descended residents; but of 22,719 deaths in Massachusetts last year, nearly 21 per cent. (4,713) were children less than one year old, while 3,192 more, of ages between one and five years, also died.

In England and Wales, in one and a half years, ending June, 1862, as appears by an official report to Parliament, 921 children were murdered or "found dead" in ditches, ponds, etc., of which 297 were in London. The same report shows that the bodies of 5,546 children under two years of age had been made the subject of official investigation as to the cause of death. Conclusive evidence of murder was produced in about one-fifth, and all but positive evidence was found in nearly one-half. But the horror does not cease here. In one year 3,000 children were burned to death under circumstances that led to the belief that they had been designedly placed where there would be a liability to such an occurrence. To these must be added those

cases where, by systematized cruelty or more caution, life is gradually destroyed, and the charge of
murder is escaped by the return of death from debility, marasmus, etc.; or where the not unfrequent
accident of being smothered in bed occurs. This
latter method of disposing of children is so common that, in some countries, it is now a penal offense for a mother to allow an infant to occupy the
same bed with herself.

In 1863, in India, 27 persons were condemned
for being concerned in cases of criminal abortion.

In the State of Ohio, during the year ending July
1, 1868, the total births reported were 36,647 legitimate, and 442 illegitimate. Of the homicides reported, one-fifth were infanticides.

The following table we have compiled from facts
gathered in the year 1848, to show the comparative
morals of the women of the *cities* of Munich, in
Bavaria, Vienna, in Austria, Paris, in France, and
London, in England. The result is startling; but
whether true now, twenty-two years later, we have
not the means of knowing:

	Munich.	Vienna.	Paris.	London.
Whole number of births	3,464	19,241	32,324	78,300
Legitimate	1,762	8,861	21,689	75,097
Illegitimate	1,702	10,380	10,635	3,203
Proportion or per cent. of illegitimate	49.10	53.84	33.16	4.1

The *countries* are far less impure than their chief
cities. By the most recent statistics before 1867, it

appears that the proportion of illegitimate births in the countries named below, where the most accurate registrations have been kept, is about, in

England and Wales...6.5	Prussia................ 7.1
Sweden..................6.5	Scotland (1858)...... 8.8
Norway..................6.6	Denmark.............. 9.3
Belgium 6.7	Hanover 9.8
France...................7.1	Austria 11.3

In England, illegitimacy is more prevalent in manufacturing and agricultural towns, and less so in seaport towns—being in London only 4.0 per cent., in Liverpool, 4.5, while in some interior towns and cities it ran up to 8, 9, 10, and even 11.5 per cent.

By the last census of England, there was a total excess of women over men of over half a million (513,706). A very large proportion of the women of England earn their own bread—three millions out of six work for subsistence. Of these, in 1861,

        1,071,201 were domestic servants.
          286,298 "   milliners.
           76,015 "   shirtmakers, etc.
          166,442 "   washerwomen.
           65,273 "   charwomen.
          812,439 "   workers *in* or *for* factories.
           84,738 "   school-teachers.
           85,798 "   boarding-house keepers.
        ─────────
        2,648,204

Most of the remainder (say 350,000) were engaged in agriculture—farmers' wives, daughters, and servants.

Illegitimacy is always a result of thus collecting females in mixed labor away from their homes. Connected with the vice that leads to illegitimacy is the awful prevalence of infanticide and abortion. By far the greater proportion of infants, over whom coroners' inquests were held, were illegitimate, and cases of abortion were alarmingly frequent.

Early marriages have been found, in all civilized countries, to be the greatest safeguard against the vice that leads to prostitution and illegitimacy, and the crimes of abortion and infanticide. The statistics confirm the natural expectation, that where the supply of women is small, the demand is great, and they marry young; and where the women outnumber the men, the reverse is the case. In 1853, about 18 per cent. of the women married in England were under age (21 years); in 1864, and again in 1866, the per cent. rose to 20, or one-fifth. Able writers have insisted, and statistics confirm the idea, that "the life of the *family* must be maintained if we would give the greatest check to excesses and crime. The mother must retain her place in the household if the household is to be comfortable; the children properly nursed; the food and clothing sufficient and nourishing; the husband attached to his family, his garden, and his home. The working class need to be convinced that the welfare of the family will be more promoted by a mother at home, than by her adding a few shillings a week to the family income—at the cost of diseased and dying children, of unfit food, of discomfort to all, and of expensive habits in an absent husband. Education and domestic training must be attended to. An ill-educated girl falls readily into temptation; an ill-educated wife can neither make nor mend clothes, nor cook, nor nurse her children properly."

The neglect of maternal duties by fashionable mothers is another prominent cause of the waste of infant life; where mothers refuse to nurse, except for a short time, their own children, and expose them to lingering but almost certain death from the use of the bottle or other food at once not natural nor nourishing. The ignorance and silly pride of mothers, in leaving their helpless little ones with bare arms, and legs, and low-neck dresses, sends many a lovely babe to an early grave. They do not stop to think that in the same dress, or want of dress, they themselves would shiver and suffer with cold, and probably have an attack of sickness.

CHAPTER XII.

PARTURITION WITHOUT PAIN OR CONSCIOUSNESS.

AT a meeting of the Obstetrical Society of Edinburgh, in October, 1862, Dr. George Smith, of Madras, in Hindoostan, communicated the following example of this:

"Some years ago I was engaged to attend an English lady during her approaching confinement, and was startled one day by a hasty summons, coupled with the information that the child had been suddenly born without warning of any kind. On reaching my patient's residence, I found that the child had been born about ten minutes, and that it was still lying, with the umbilical cord uncut, close to the mother's body. The native female servant, at the lady's order, had left the child untouched, merely raising the bedclothes a little to permit the free access of air for the purpose of respiration.

"On inquiry, the lady informed me that she had been for some time expecting her confinement daily; that the previous night she had felt as usual; but that she had had occasion to rise frequently to attend upon her sick child, and that she had got up as usual about half-past five A. M., feeling well, and having no indication of the near approach of labor. Further, that during the forenoon she had walked down a long flight of steps, and across a graveled walk to a smaller house within the inclosure of her own grounds, where, feeling a little tired, she had lain down upon a bed; that soon after she experienced slight discomfort, likened by her to ill-defined uneasiness of the abdomen under the operation of a mild laxative, followed by an impression that some solid warm body was lying in contact with her person; that she directed her servant to look below the bedclothes, and that the attendant, on doing so, found to her surprise the child entirely extruded.

"My patient assured me repeatedly and earnestly that she was quite unconscious of the whole parturient process culminating in the birth of the child, and expressed herself both surprised and alarmed at a delivery so painless and instantaneous. As she was hourly expecting her delivery, it is but reasonable to suppose that she had been for some time acutely alive to the earliest intimations of commencing parturition, and it is surely remarkable that nothing occurred from which she could have suspected that the act had actually commenced. My patient had no object in deceiving me, and I am quite satisfied of the entire truthfulness of her often—to me—repeated statement.

"This case has a medico-legal significance, as well as a practical. If a female, awake, in perfect health,

in the exercise of sound reason, and hourly expecting
her confinement, having no object for its concealment,
but many reasons for its occurrence being welcomed
by her friends, can be the subject of painless, unconscious labor, preceded by no appreciable premonitory
symptoms, and making itself known only when the
extrusion of the child has been completed in the way
described, how much more may we be inclined to
yield belief to cases in which it has been averred
that delivery has taken place during sleep, without
waking the mother, and to others, in which it has
been maintained that, owing to the painlessness of
the parturient process, the child's life has been lost
by a fall on the ground, or by being engulfed in a
latrine? The child was a female, small, but not much
undersized. The mother's first labor—this was the
second—was a normal one, accompanied by the usual
signs, and extending over six hours in its duration."

In the discussion which followed, Dr. Pattison stated
that he had once attended a primiparous patient who
suffered no pain at all during labor. He had not been
summoned to the case, but happened to call at the time;
the child was born quite easily, the patient only ex-
periencing a feeling of pressure.

Dr. Wilson had once been called to see a woman
who had been delivered without any pain, whilst she
was walking about in the house; and he found the
child lying on the floor with the umbilical cord torn
across.

Dr. Cochrane thought that such a case as that

related by Dr. Smith might more readily occur in a warm country with a relaxing climate. But he had himself seen a woman who had just been delivered of a child almost unconciously, as she was getting out of bed.

Dr. Andrew Balfour stated that he had attended, when in China, the wife of an engineer on board a steamer, who suffered from remittent fever in the eighth month of her pregnancy. The whole ovum in that case was expelled entire without any warning; and when he (Dr. B.) arrived and ruptured the sac, the fœtus was already dead.

Dr. Pattison said Dr. Thatcher used to tell his class of a case where he found the patient had been delivered of an entire ovum with unruptured membranes. Dr. T. had been summoned by the husband, who was in great dismay, because, as he averred, his wife had given birth to a " leg of mutton."

Dr. Alex. R. Simpson stated that Von Ritgen, the venerable professor of midwifery at Giessen, had told him, that in a long course of his practice he had met with no less than seventeen cases of labor where the patient had experienced none of the ordinary labor pains; and he (Professor Von Ritgen) had been led to form the conclusion that in perfectly natural labor, pain should not necessarily be experienced, and that we had come to regard pain as a natural and necessary concomitant of labor, merely because women were almost never in a perfectly healthy condition when we are summoned to aid them during parturition. He (Dr. A. R. S.) thought that if Professor Von Ritgen's position could be established—and the facilities of parturition among savages went far to prove its truth—then the objection sometimes made to the use of chloroform in labor, on the ground of its being contrary to nature, would be most completely done away with.

The medical journal which reports the above, contains references to twenty-two other cases, where the patients had experienced none of the ordinary labor-pains.

A case is reported in Braithwaite's *Retrospect*, to which Mr. Rawson was called immediately on the waters being discharged. No pain was present, and the patient was asleep on his arrival. On examination, he found the *os uteri* dilated, and the head presenting. The child was slowly and *unintermittingly*, but forcibly, expelled. She betrayed no symptoms of uneasiness whatever; and, though he watched her countenance, she did not exhibit the least consciousness of the child's expulsion, but expressed her surprise on seeing it. The child was strong and lively, and, with the mother, did well. The mother was about twenty-two years old, short, plethoric, and *healthy*.

Dr. Montgomery, in his work on Pregnancy, doubts the possibility of such an occurrence, excepting under peculiar circumstances, *certainly not in a first delivery.*

Dr. Beck, in his book on Medical Jurisprudence, says the possibility of a woman being delivered without being conscious of it, is disbelieved, excepting some extraordinary and striking cause intervened.

A case of sudden and unconscious delivery is reported by Dr. John Shortt, page 210, of *Trans. of Obst. Soc.* of London, for 1862. A Hindoo woman of caste, whilst walking with two companions—all three carrying loaded baskets upon their heads—suddenly, without any preceding pain or other premonition, felt her child slip between her thighs to the ground; the after-birth came away soon after. She was in good health, twenty-eight years old, and the mother of two living children; knew herself to be pregnant, but did not expect to be confined for some days.

The wife of a soldier arrived at Washington city on the eve of June 1, 1863, after traveling day and night from the northern part of the State of New York. During the night she visited the water-closet, when the pangs of labor came upon her; and before she could arise from her position, the whole contents of the womb had disappeared through the opening in the seat. When the physician, Dr. Robbins, arrived, he heard the cries of the child in the drum, five feet below the surface of the earth. It was soon rescued, and proved a well developed male child, weighing eight pounds. The mother was a small woman, with an unnaturally large, rigid pelvis. She had previously given birth to two children, with but little pain, and the deliveries were soon over.

CONGENITAL MONORCHIA IN MAN.

The *Medizinische Jahrbücher*, for 1868, says that Dr. Gruber, of St. Petersburg, in searching the literature of the last three hundred years, has found but twenty-two genuine cases of congenital deficiency of one testicle (monorchia). From a careful consideration of the details of these cases, and also one observed by himself, he is enabled to assert that the subjects of this congenital deficiency are generally well-developed, free from other malformation and structural defects, and capable of arriving at an advanced age. The scrotum is usually well developed, and the testicle is more frequently absent on the right side than on the left. The seminal apparatus and genital organs on the opposite side rarely present any associated anomalies. The vesicula seminalis corresponding to the affected side is generally normal in size and form, and still acts as a secreting gland. No spermatozoa, however, have been found in the fluid taken from the vesicula seminalis on the side of the monorchia, although these bodies are generally present in great numbers in that of the opposite gland. The subject of a congenital monorchia, so long as the opposite testicle is well developed, is not incapable of procreating. In a case reported by Graaf, the man was the father of four children.

A very earnest and long-continued discussion as to the extent of medico-professional secrecy—"*le secret medical àpropos de marriage*"—originated in

the following melancholy fact : An eminent professor, M. Delpech, having heard that the daughter of
an intimate friend was about to marry a young man
who he knew had but one testicle, for *her* sake (?)
informed the father, who broke off the match and
forbade the suitor further addresses to his daughter.
The thwarted lover did not rest until he obtained
vengeance by assassinating Delpech in his carriage,
with his coachman, and then took his own life—a
triple murder. As a *monorchide* simply, he was not
diseased, and no unworthy candidate for matrimony.

A gentleman informs the writer of this that, when
at school, in Greensburg, Indiana, about the year
1853, he well knew a boy whose left testicle was
mashed and flattened by a fall when about the age
of six, so as to permanently prevent its natural
growth or enlargement, thus incapacitating it for
the usual healthy secretion. When the boy, H., had
grown to manhood, and been married some six
years (the last heard of him), he was the father of
two boys and *no* girls. The writer regards it a
physical impossibility for him to beget female children, because he was practically a monorchide.

EMASCULATION.

Dr. D. N. Rankin, physician to the Western
Penitentiary, of Pennsylvania, reports in the *Am.
Jour. of Med. Sci.*, for July, 1867, the case of a pris-
oner named W. J. Davids, who, on May 6, 1866,

with the knife ordinarily used to cut his meat and bread, after sharpening it on the stone floor, deliberately removed both his testicles. Eight months afterward he was in excellent health, but considerably troubled with obesity, and said he was much better contented to remain within his solitary cell, where he sees no person except the officers of the prison. The reason given for the barbarous act was that he had been terribly tormented, during his sleeping hours, for some considerable time, with annoying lascivious dreams which became insupportable. He removed his testicles because he supposed them the cause of all this trouble. His recovery was sufficient in three weeks to enable him to resume his usual work in the prison.

A man who, in 1867, was farming near Shelbyville, Indiana, had been a great favorite with the girls and very fond of their society up to about his sixteenth year. About this time he had a dangerous attack of mumps, complicated with a bad cold, causing the mumps to settle in his testicles; and on consultation he was castrated as the only means of saving his life. His female friends immediately forsook him, and he remains a bachelor to this day. Their neglect of him was so pointed that it became a proverb. From being thin and consumptive-looking he rapidly became very fleshy, in striking contrast to most of his relatives, who were thin and spare.

The Cincinnati *Lancet*, December, 1868, records a case of double castration for epilepsy, in 1861, by

Dr. J. I. Rooker. The patient, a young man, had but one attack of epilepsy afterward, and that on the night after the operation. He had been in the almost daily habit of masturbation, for eight years, each indulgence followed by epilepsy. Previous to the operation he was not able to do a day's work, owing to general debility and loss of mind. Afterward he served three years in the army, and in 1868 his weight, from one hundred and twenty increased to one hundred and sixty pounds, his "nervousness" had all left him, he could do as hard and as much labor as any man, his intellect was good as ever, his voice had not changed, and he said he had but little "passion left for the women."

Seven other successful cases, of a like character, are reported in the United States.

REMOVAL OF OVARIES AND UTERUS.

The removal of *one* ovary is no longer a rare operation in surgery, having been successfully and frequently accomplished, within the last fifteen years, by every extensive ovariotomist—not, however, except in remarkable cases, being the cause of the operation, but incidental and necessary in the excision of an ovarian tumor. The removal of *both* ovaries has very seldom been found necessary.

Dr. S. Choppin, in the *South. Jour. Med. Sci.*, February, 1867, relates a case of a woman aged thirty-eight years, suffering with a tumor six inches long by three and a half inches in breadth. On the 19th January, 1861, with the patient under the influence of chloroform, he removed the uterus, left Fallopian tube, and left ovary. One month after she was presented to the class of the New Orleans School of Medicine with her womb in her hand. She recovered, became a robust and healthy woman, and lived more than three years.

It is stated in the French *Journal*, 1863, that Dr. Kaberle, of Strasburg, in opening the abdomen for the removal of a fibrous tumour of the uterus, found this organ and one ovary so extensively diseased that he removed both, leaving only the neck of the uterus. Five weeks after the operation the patient is reported to be convalescent.

The *Western Lancet*, for 1852, p. 182, records that Dr. —— extirpated, by a peritoneal section of nine inches, from Fanny Gould, an ovarian cyst consisting of a hypertrophy of the left ovary, etc. During the winter after the catamenia appeared regularly, and in April, 1850, she married. In January, 1851, her menses ceased, and 282 days after, on October 9, 1851, she was safely delivered of a male child, weighing seven pounds.

Dr. E. R. Peaslee, in the *Am. Jour. Med. Sci.*, April, 1851, reported a case of double ovariotomy; patient twenty-four years old, unmarried: "She married about a year after; has, of course, never menstruated since that time, nor conceived. She has enjoyed uniform good health, never having had even

13

headaches, periodical or otherwise. With the *two* exceptions above specified, she is capable of fulfilling *all* the functions attributable to her sex. Neither her external physical conformation nor her mental characteristics have undergone any change in consequence of the absence of the ovaries. She is in all respects now, at the age of thirty-six, a splendidly developed woman."

Another case of double ovariotomy, in the practice of the same eminent surgeon—recorded in same journal, April, 1863—was that of Mrs. S., of Vermont, a highly educated lady, of delicate constitution, thirty-five years old; eleven years married, but had never conceived; had had an attack of inflammation of the left ovary one year after marriage, and two years subsequently was treated for ulceration of the cervix uteri. From January, 1861, to August 21, 1862, she was tapped twenty-six times, discharging each time from twenty to thirty pounds of fluid. A portion of the time the tappings were only twelve to fourteen days apart. Menstruation continued regularly until it ceased entirely, in May, 1862. The right ovary was first removed, and another tumor found whose removal took the left ovary. Recovery slow but perfect, and seven months after she was in excellent health.

A GENUINE HERMAPHRODITE.

Dr. Avery gives, in the *Med. and Surg. Reporter,* 1868, the details of a case which he declares to be a genuine hermaphrodite. She was a native of Nova Scotia, unmarried, twenty-four years of age, five feet ten inches high, with a deep, coarse voice, masculine frame and face, and all the characteristics of an ordinary coarse woman. The mammæ were undeveloped; the clitoris, resembling a penis in flaccid state, was two inches long and half an inch in diameter, with well developed gland and foreskin. No orifice was discovered. A vagina two inches deep, and well formed, existed, but a close examination per rectum and bladder could not discover any trace of a uterus; the meatus urinarius and vestibule were perfect; the right labium major was quite natural, and of usual size; the labia minora were traceable, but in the folds of the left labium there appeared a large pendent *tumor* resembling the left *testicle* of a man, with a well developed scrotum of usual size, of some four inches in length, resembling in every respect the scrotum. Tracing what appeared to be the cord up, he found it made its exit from the external abdominal ring, and having every indication of a spermatic cord; the epididymis appeared to be natural; in fact, every thing resembled a *testicle.* Her object in coming to Dr. A. was to have the tumor removed, as it annoyed her. She had noticed nothing of it until she was ten years of age. After removal, the tumor was examincd with a powerful microscope, by three physicians, when cellular structure and convoluted tubes were visible, with rudimentary spermatozoa; in fact it was declared a *testicle*.

This is the only case, he believes, on record, where a *testicle* has been discovered in a *woman*. The *fact* can now be settled, that such a thing as a hermaphrodite has existed.

OVARIOTOMY.

The formation of the female abdomen is such, that the organs will still perform their functions, notwithstanding the presence and continued growth of a tumor until it reaches the size of the womb at full term of pregnancy. A surgical operation to remove such a tumor is usually not necessary even until its weight exceeds fifteen or twenty pounds. It sometimes happens that the growth of such a tumor, as steady in its enlargement almost as the fœtus in the womb of a pregnant woman, gives occasion to the tongue of slander to vilify the character of a virtuous sufferer. The knife of the surgeon, successfully applied in cases of ovarian dropsy or tumors or cysts, promptly removes the cause, but can never efface the remembrance of the slanderous gossip, which often aggravates and intensifies the sufferings of the patient. These tumors are of all sizes and weights, from those so small as to cause but little personal discomfort, up to eighty, one hundred, and as high as one hundred

and fifty pounds. One of this size was removed from
Mrs. M. A. L., of Albany, N. Y., in May, 1867.
She died from the operation, the performance of
which she strenuously opposed, although she had
become a burden to herself and a wonder to her
friends. For an interesting report of several cases
of successful double ovariotomy, see *ante* pp. 193–4.

Mr. T. Spencer Wells, of London—the most celebrated surgeon in the world, in this class of cases—
completed, in Feb., 1868, the operation in 250 cases,
in just ten years, with a result of 180 recoveries and
70 deaths—a mortality of exactly 28 per cent. His
success had constantly increased, and the mortality
been steadily reduced—for of the last 50 cases only
eight died while 42 recovered, a mortality of only 16
per cent. He confidently expected still more favorable results. In March, 1870, he completed 50 more
cases, of which 15 died and 35 recovered—a mortality
exceeding his experience except in his first 100 cases.
From extended observation he concluded that previous
tappings do not increase the mortality, and may be
useful—while the danger from tapping is small.

Next to Dr. Peaslee and Dr. Atlee, probably the
most successful—if not the most extensive—operators
in ovarian diseases, in the United States, is Dr. Alex.
Dunlap, of Springfield, Ohio, recently President of
the Ohio State Medical Society. Of 51 cases in his
hands, between the years 1843 and 1870, 11 died,
but several of these were not properly attributable to
the operation; latterly, his success has been remark-

able; the tumors removed ranged in weight from 14 to 136 pounds.

A table was prepared by Dr. W. L. Atlee, in 1851, of 222 cases, being all the known operations of ovariotomy from 1701 to 1851. Of that list, the oldest person subjected to operation was in 1844, Dr. Atlee's first patient, aged 61 years. In 1852, he operated successfully on a woman 68½ years old. His youngest patient, and successful, was a young woman of 16, in 1862. In 1846, Dr. Hawkins operated on a young woman of 18, at that time the youngest subject known.

The following table is by no means full, but will be interesting to the scientific reader, as showing the progress made in this line of surgery:

Operator.	Cases.	Recoveries.	Deaths.
T. Spencer Wells..............	300	215	85
Dr. Alex. Dunlap..............	51	40	11
Dr. J. Clay.....................	104	72	32
J. B. Brown...................	19	13	6
Dr. W. Tyler Smith............	20	16	4
Thos. Bryant..................	10	6	4
Cases collected by Dr. J. Clay, up to Jan. 1, 1860............	425	242	183
Cases collected by Dr. Peaslee, between 1860 and 1854.....	150	99	51
	1079	703	376

Of the recent cases of experienced operators, over 82 per cent.—in some hands still more—have been saved by the operation.

OVARIAN PHYSIOLOGY AND PATHOLOGY.

In an excellent volume entitled "Contributions to assist the Study of Ovarian Physiology and Pathology," London, 1865, Dr. Charles G. Ritchie maintains that active changes are incessantly taking place in the ovary from the earliest to the latest period of life; that both before and after puberty, during pregnancy and lactation, during any suppression and even after the final cessation of the menses, vesicles are constantly being formed in the ovary, and constantly perishing. He also claims, that ovarian cysts may be produced in many different ways, each morbid process being a modification of some physiological action occurring normally in the ovary; that these cysts are occasionally "moles" or ova which have undergone a certain amount of development; and that dermoid cysts are certainly derived from an ovum, and have little in common with that form of subcutaneous dermoid cyst not unfrequently met with in the region of the eyelids.

The following conclusions arrived at by the father of Dr. Ritchie, in the course of his researches, are also strongly advocated by his son:

"1. The Fallopian tubes of the infant, and of the child before puberty, are perfect in their structure, although their patency is more or less obstructed at birth by the presence in them and in the uterus of a tenacious glue-colored mucus.

"2. The ovaries of new-born infants and children are occupied, sometimes numerously, by Graafian vesicles or ovisacs, which are highly vascular as early as the sixth year, and vary in size from the bulk of a coriander-seed to that of a small raisin, in the fourteenth year, at which time, also, they are filled with their usual transparent granular fluid.

"3. The Graafian vesicles contained in the ovaries prior to menstruation are found, as they also are in every other period of life, in continual progression toward the circumference of the glands, which they penetrate, discharging themselves by circular-shaped capillary-sized pores or openings in the 'peritoneal coat; the presence of the catamenia being thus no indispensable prerequisite to their rupture.

"4. The establishment of menstruation does not necessarily give rise to any immediate modification in the manner in which the ovisacs are discharged, or in the subsequent change which these bodies undergo; but in some cases the conditions which obtain in the period before puberty are extended, for a time, into that of menstruation.

"5. The ovisacs of the human female do not require the establishment or presence of menstruation for their development or rupture. Vesicles of adult size may exist and be discharged about the age of puberty, as also at other periods of life, independently of menstruation; and this state may be present in its normal form for at least eight consecutive periods, without a vesicle being ruptured, unless after the manner, and with the phenomena which occur in childhood.

"6. In early infancy, extreme old age, and long-continued organic disease, the ova are minute, transparent, and structureless cells; and in advanced childhood, soon after the critical age, and during pregnancy and lactation, they are more or less organized, larger, and in the latter state are often so well maturated, that about *one-third of the renewed pregnancies of married women take place while they nurse.*

"7. In children and others in the circumstances now mentioned, the exercise of organic power which occasions the secretion and extrusion of ova, is attended also by that of an opaque mucous fluid in the tubes and uterus; but on the attainment by the female of maturity, and onward to the period of critical life (when the animal powers become again diminished, and the caliber of the arteries reduced), the ovarian orgasm, like many other vital actions, undergoes periodical augmentations of power, during which, unless when prevented by the disturbing influence of other functional processes or by disease, it extends to the nervous and vascular tissues of the uterus, and gives rise to the formation within this viscus of decidual vessels, which exude lymph and red globules, the latter being evacuated, mixed with watery fluid, mucous matter, and some of the salts of the blood, in the form of menstruation.

" 8. That the periodical congestion or increase in size of the ovaries before and at the usual menstrual terms is not caused by the presence or bursting of ripe vesicles, is plain from the consideration that the ovaries are often crowded with such, some of which also occasionally give way, and empty themselves by capillary sized openings in the surface of the glands in women who have never menstruated; and in others who have ceased to menstruate, one and two large and fully-developed vesicles are frequently seen.

" 9. The elimination of ova, and the process of menstruation, are correlative effects of the vital powers of the ovaries (just as the secretion of mucus, and of gastric juice, and the chymification of the blood are those of the stomach); and to suppose the rupture of the Graafian vesicles to be the cause of the menses, is to mistake a frequent association and, to some extent, effect for a uniform cause.

" 10. The principal use of menstruation is, in regard to the ovary, to provide an accessory by which the maturation of its vesicles, and the absorption of their peritoneal and tenacious proper coats, and their extension generally, may be promoted throughout the child-bearing period of life; and, in reference to the uterus, to furnish a nidus within its walls by which the ovum may be entangled, retained, and nourished."

YEARLY RIPENING OF OVA IN WOMEN.

Dr. Mattei, in *Gaz. des Hôp.* 22, 1859, " regards the theory of the monthly maturation of ova concurrently with menstruation as erroneous; and believes that for each ovary only one annual ripening takes place. January, February, March, and April are the months most favorable for this. The appearances of this condition come on sometimes very gently, at others very painfully. The general appearances are alteration of voice, sleeplessness, at times neuralgias, prostration, vomiting, frequently palpitation, cough, hoarseness, without material changes in the breasts. As local symptoms there are—sensation of weight or pain in the abdomen, from the sacrum to the thighs, and especially pains on that side of the pelvis on which the lymphatic glands are swollen and tender; there is also heat and excitement of the external genitals; the menstruation is disturbed—it is seldom rendered more profuse, frequently more scanty, coming on earlier, and attended with nausea, at times leuchorrhœa, diarrhœa, dysuria, sympathetic symptoms in the breasts; the excitation of the ovarian region causes pain, nausea, even hysterical cramps; hematocele, peritonitis, and phlegmon may occur, according to the individual; these symptoms may last for four, twelve, or twenty-eight days, and disappear altogether, or pass into symptoms of pregnancy or false conception; the interval between the ripening of the ova in the two ovaries is variable—

the minimum observed was four days, the maximum five months." Dr. Mattei further says that this yearly ripening mostly ceases at the same epoch as the germination of plants, and the rut of animals.

CONNECTION OF THE FALLOPIAN TUBE WITH THE OVARY.

Dr. Panck claims that the adhesion of the orifice of the Fallopian tube with the ovary at the time of conception, takes place through a newly-formed membrane, which at a later period is absorbed. He examined the body of a girl who had died shortly after conception, and found a delicate new membrane, which fastened the fringes of the tube to the ovary. In subsequent researches, he found a similar membrane so frequently as thirty-four times out of fifty-eight cases, but only in women who had borne children.

CHAPTER XIII.

DISPROPORTION OF THE SEXES.

EMIGRATION has a remarkable influence in effecting a disparity in the sexes—the males largely predominating in the new States and Territories, and the reverse generally in the old settled States.

The most recent census returns show that the males of California outnumber the females nearly sixty-seven thousand, or about one-fifth of the population. In Massachusetts the females outnumber the males sixty-three thousand. Connecticut, seven thousand. Michigan shows nearly forty thousand excess of males; Texas thirty-six thousand; Wisconsin, forty-three thousand. In Colorado, the males are as twenty to one female. In Utah, the numbers are nearly equal; while in New York there is a small preponderance of the females; the males are most numerous in Pennsylvania. In the District of Columbia, in a total population of 118,-867, there are several thousand females more than males.

The female population of England and Wales

exceeds the male by 513,706, as ascertained by the census of 1861. But the male births exceed the female ones—the proportion in 1853 being 105.1 boys to 100 girls; in 1860, 104.7 boys to 100 girls; and in 1864, 104.2 boys to 100 girls. This excess of males is reduced by (1.) Infant boys die in greater numbers than infant girls; (2.) The more dangerous and exposed occupations of men cause greater proportional mortality; (3.) Many sailors are nearly always absent from the country; (4.) Men in greater numbers emigrate in search of fortune. Thus, the proportion of population is almost exactly in reverse to that of births; for, in 1861 there were 105.25 women to 100 men.

In the city of Philadelphia, in 1861, the excess of male births was nine per cent., or as 109.7 males to 100 females. In the year 1867 this excess was slightly greater (9.9). In the month of June it ran up to 17 per cent., while in May it was only 4.4, in April 7.8, and in July 9.5. In September, only three more males than females were born. The deaths among the males exceeded the females by 3.57 per cent., and this excess was nearly all under twenty years of age.

In the State of Ohio, during the year ending July 1, 1868, out of a total number of births reported of 37,089, the excess of male births was 2,329, about thirteen and one-third per cent, or as 113.3 males to 100 females. Of the deaths reported during the same year, the excess of white males over females was 18 per cent., and of black males over females 22.22 per cent.

In the State of Michigan, in the nine months beginning April 5, 1868, the number of births was 19,171, or as 112.55 males to 100 females.

In a paper, by Mr. W. L. Sargent, on the vital statistics of Birmingham and seven other large towns in England, results are arrived at, different in some respects from those commonly received. Thus, while admitting that there are more male deaths on the whole than female, the author states' that "in the first five years of life there is a large excess of male deaths; that from five to ten years of age the male excess is very small; that from ten to fifteen the male excess is decided, but at fifteen the tide turns, and until twenty the female deaths are largely in excess; that from twenty to thirty-five the male and female deaths are equal, and that after thirty-five the male deaths are again in excess." Mr. Sargent adds in a note the "singular fact," obtained from Mr. Makeham, who is such high authority in vital statistics, "that among the males of the upper classes, from the age of sixteen to that of twenty-three, the annual rate of mortality rises rapidly from about eight to a maximum of fifteen per one thousand, after which it gradually diminishes."

The number of women above the age of twenty, who *must* remain single, in consequence of the actual disproportion of the sexes in England and Wales, is between 300,000 and 400,000. The number of adult women who actually *are* single is 1,537,000, of whom 1,230,000 are between twenty and forty years of age.

MARRIAGE BY BANS.

The once popular and common mode of publishing the bans of marriage, has, to a great extent, fallen into disuse; indeed, in many of the United States, there is no provision by law for this, but a license from a court is especially required. In Ohio either mode is lawful—the statute stipulating that previous to persons being joined in marriage, notice thereof shall be published, in the presence of the congregation, on two different days of public worship, the first publication to be at least ten days previous to the marriage, and within the county where the female resides ; *or* a license shall be obtained from the probate court in the county where the female resides. The law is peculiar in requiring the license to issue in the county of the residence of the female; any other arrangement is an evasion of the law.

The number of marriages by *bans* is gradually diminishing, and will, probably, soon disappear enentirely, except where sustained by the usages of one or more religious sects of limited membership. In Ohio, during the year ending July 1, 1868, out of 28,231 marriages, only 460 were by bans, or 1.63 per cent.

In the State of Kentucky, license from the county court is required in all cases, except where either party belongs to a religious society, having no officiating priest or minister, whose usage is to solemnize marriage at the regular places of worship, and by consent given in the presence of the society. Kentucky is more liberal than Ohio, in providing that, if the female be of full age or a widow, upon proper application, license to marry may issue any where in the State.

DIVORCES.

The looseness of the divorce laws of several of the United States, the great number of causes for which divorces are granted, and the extended system of fraud by which citizens of other States are encouraged to avail themselves of these laws, are all indications of a lamentable state of public morals. The increasing tendency to apply this sovereign balm for family jars and uncongenial marriages, so apparent in some of the States, does not exist in others. In Ohio, the only State whose statistics are at hand, there was a positive decrease in the number of divorces granted from the last year of the war, 1865, up to and including 1868, four years. During the year ending July 1, 1868, the number and causes of the divorces granted were as follows:

Adultery	233
Cruelty	114
Absence and neglect	365
Drunkenness	66
Impotency	1
Imprisonment in penitentiary	15
Fraud	9
Wife or husband living at second marriage	6
	809
Five counties estimated	38
Total in the State	847
Number of divorces in the year ending July 1, 1867	975
Decrease	128

The following table shows the number of divorces and their causes during four years:

	1865	1866	1867	1868	Total.	Per Ct.
Adultery	254	389	317	233	1193	30.48
Absence and neglect	230	394	417	365	1406	35.93
Cruelty	199	134	132	114	579	14.79
Drunkenness	48	79	73	66	266	6.80
Fraud	18	30	10	9	67	1.71
Miscellaneous	183	133	26	60	402	10.27
	932	1159	975	847	3913	99.98

In 1865 the divorces stood in proportion to the number of marriages, as 1 to 24, nearly; in 1866, as 1 to 26; in 1867, as 1 to 30; in 1868, as 1 to 33.

THE SEASONS—MARRIAGES.

The season of the year at which most marriages take place it is not at all important to know, but all

14

statistics concerning social life are interesting. In England, during the ten years from 1856 to 1865, inclusive, the average number of persons married per year was nearly 17 (1.673 per centum) to the 1,000 of population. The average per centage, per quarter, of persons married was:

In January, February, and March.........1.398
" April, May, and June....................1.698
" July, August, and September...........1.621
" October, November, and December....1.981

The quarterly average is 1.678 per cent. for the whole period. In the first quarter of the year 1865, nearly one-fifth (19.8 per cent.) of marriages took place, in the second and third quarters nearly one-fourth in each (24.7), and in the last quarter a little less than one-third (30.8) out of each 100. This proportion was remarkably uniform during the seven years from 1860 to 1867.

In the State of Michigan, in the nine months of the year 1868, from April 5th to December 31st, there were 5,485 marriages, distributed through the months and seasons as follows:

First Quarter.	Second Quarter.	Third Quarter.
April..........569	July.............635	October..........738
May............611	August..........533	November......662
June...........487	September......718	December.......524
1,667	1,886	1,924

THE SEASONS—BIRTHS.

The population of England, Ireland, and Scotland, in the year 1866, was, in round numbers, 30,000,000. During the preceding ten years the average birth-rate was 34 to the 1,000 (34.83) ; the births in the first half of years being 3 to the 1,000 greater than in the last half. The actual number of births in England,

For the first quarter of 1865, was........194,287
"   second   " 192,921
"   third   " 181,642
"   fourth   " 179,020

Total births in 1865......................747,870

In the State of Michigan the returns for the last three quarters of the year 1868, show as follows out of 19,171 children born alive :

Second Quarter.	Third Quarter.	Fourth Quarter.
April.........1,607	July...........2,163	October..........2,246
May...........2,037	August.......2,399	November......2,099
June.........1,944	September...2,362	December.......2,283
5,588	6,924	6,628

In a record of 2,387 cases of obstetrics, kept by Dr. H. Corson, of Montgomery County, Pennsylvania, the births were as follows :

Winter.	Spring.	Summer.	Autumn.
December....222	March....221	June.....199	September...211
January......190	April......170	July......213	October......180
February.....227	May.......183	August..207	November...203
639	574	619	594

In a similar record of 1,000 cases, kept by Dr. M. C. Richardson, of Hallowell, Maine, the births took place as follows:

Winter.	Spring.	Summer.	Fall.
December......62	March.......99	June......81	September...92
January.........78	April........87	July.......78	October.......97
February.......85	May 87	August ...92	November...62
225	273	251	251

The following additional statistics, gathered by the same systematic physician, will be read with curious interest by mothers:

There were delivered between the hours of—

12 and 1 A. M......35 Cases.	12 and 1 P. M......27 Cases.	
1 and 2 " ......44 Cases.	1 and 2 " ......25 Cases.	
2 and 3 " ......58 Cases.	2 and 3 " ......29 Cases.	
3 and 4 " ......56 Cases.	3 and 4 " ......29 Cases.	
4 and 5 " ......72 Cases.	4 and 5 " ......39 Cases.	
5 and 6 " ......62 Cases.	5 and 6 " ......26 Cases.	
6 and 7 " ......51 Cases.	6 and 7 " ......29 Cases.	
7 and 8 " ......47 Cases.	7 and 8 " ......36 Cases.	
8 and 9 " ..... 43 Cases.	8 and 9 " ......44 Cases.	
9 and 10 " ......52 Cases.	9 and 10 " ......28 Cases.	
10 and 11 " ......47 Cases.	10 and 11 " ......46 Cases.	
11 and 12 " ......30 Cases.	11 and 12 " ..... 45 Cases.	

Whole number of cases between midnight and noon......... 597

"   "   "   "   noon and midnight......... 403

Cases terminating between midnight and 6 A. M............... 327

"   "   "   6 and 12 A. M........................ 270

"   "   "   noon and 6 P. M..................... 175

"   "   "   6 P. M. and midnight.............. 228

"   "   "   6 P. M. and 6 A. M................. 555

"   "   "   6 A. M. and 6 P. M................. 445

Of the 1,012 children born, there were :

Of the first child of the		Of the seventh	37
mother	291	Of the eighth	18
Of the second	220	Of the ninth	18
Of the third	175	Of the tenth	15
Of the fourth	110	Of the eleventh	9
Of the fifth	69	Of the twelfth	2
Of the sixth	46	Of the thirteenth	1

" The mother delivered of the thirteenth child was one of the two delivered of the twelfth ; and both of these women have given birth to their fourteenth child. *One* of them lost all her first ten children before they were eighteen months old. The eleventh, twelfth, and thirteenth are now alive and healthy ; parents intemperate. The other family is sickly ; parents scrofulous ; and half the children have died before the age of ten years.

" The longest time after marriage, before the birth of the first child, was fifteen years; the *shortest* time fifteen *minutes.* The oldest mother of first child, forty-six years ; the youngest mother of first child, fifteen years and nine months. The longest interval between the births of two consecutive children, nineteen years ; the shortest, ten months and twenty-nine days, the mother being delivered of a living child March 8th and February 6th following. June 11th and December 11th of the same year, Mrs. —— procured an abortion, being each time from two to three months pregnant.

" No case is reckoned among these enumerated

where the delivery occurred previous to the fifth
month. Nine hundred and sixty were confined at
the expiration of the full term of gestation. Forty
cases were premature; in twenty-five of them the
child was still-born, and in fifteen, born alive.
Twenty of the children delivered at the full term
were still-born.

"Nine hundred and eighty-eight were delivered
of a single child, and twelve of twins.

Whole number of children born	1012	Children born in wedlock	1000
There were of boys	551	Illegitimate children	12
Of girls	461	Unmarried, first child	9
Excess of boys over girls	90	" second child	1
Still-born	45	Having husbands	2

"Of the twins, two pairs were boys, three girls,
and seven a boy and a girl. One pair born *February 29th*. (When will the anniversary of their birth
day occur?)

"The average duration of labor, ascertained as accurately as possible from the patients themselves
and others, has been eleven hours and thirteen
minutes. The longest labor, with regular pains,
ninety-six hours; the shortest, one single pain. The
average time of attendance before the birth of the
child, three hours and twenty-two minutes. Two
of my patients have been insane. One became so
three weeks before confinement with fourth child.
A perfect recovery on delivery. In the other case
the insanity manifested itself on the second day after

the birth of first child. Recovery at the end of four weeks. Seven of my patients have died in child-bed. Three from convulsions above mentioned; one from pneumonia, commencing three days before confinement; one from consumption; peritonitis, one; one died in twelve hours, after an easy labor with first child, from no apparent cause, expiring in ten minutes after taking part in a cheerful conversation; one other lived but a *month*, suffered from incessant vomiting and aphthæ for two months before confinement; could take but little food, and died from exhaustion.

" Cases of abortion occur too frequently in this community, and, strange as it may seem, most of them among the married, brought on in many instances *to avoid having large families.* These cases are, in general, attended with more flooding and greater prostration, and a much larger proportion of deaths than follow confinement at the full term of gestation."

CHAPTER XIV.

MENSTRUATION, HOW ESTABLISHED.

IN Chapter V, page 74 and following, we have already presented 1018 cases, designed to show the various ages at which menstruation is established ; and in the same chapter, pages 81–83, 96 cases to show the ages to which it is prolonged. Another interesting view of this distinguishing function of the female system, is the ease, or pain, or irregularity attending the establishment of menstruation.

Dr. Bennet, when physician-accoucheur of the Western General Dispensary of London—an institution in which nearly 10,000 patients annually are treated by its medical officers—very carefully prepared a synopsis of a large number of cases presenting uterine symptoms. From these, with great labor, and especially to illustrate our present subject, we have selected out 255 cases, as below. The term "*painfully*," as applied by Dr. Bennet to menstruation, means that, physiologically, menstruation was attended with considerable pain, and was anomalously scanty or abundant, frequent or rare. Whereas "*easily*" means, that it was free from any of these physiological peculiarities; and "*irregularly*," that its manifestation was irregular, although unaccompanied by marked pain. In addition to the columns below, showing the character or manner of the menstruation, we have prepared one—the first in order—showing the per cent. or proportion of the 255 cases in which menstruation was established at the ages mentioned :

Age of Menstruation.	Per cent.	Easily.	Painfully.	Irregularly.	Total.
At 10 years......... 2		3	2	0	5
" 11 " 6		8	7	0	15
" 12 " 11		10	17	1	28
" 13 " 13¾		17	17	1	35
" 14 " 16		15	23	3	41
" 15 " 17⅓		22	20	2	44
" 16 " 10⅔		18	8	1	27
" 17 " 8⅓		8	10	3	21
" 18 " 9		11	8	4	23
" 19 " 4		4	6	0	10
" 20 " 2		3	1	2	6

The age at which menstruation ceased he recorded in only 11 cases—1 at 43 years, 2 at 44, 2 at 47, 1 at 48, 1 at 49, and 4 at 50 years of age.

Of the 300 cases from which we gain the above results, 6 were below 20 years of age, 31 between 20 and 25, 59 between 25 and 30, 117 between 30 and 40 years, 57 between 40 and 50 years, 21 between 50 and 60, 3 were 60 years of age, one was 63, one 65, and one 70 years of age, and of 3 the age is not given.

PRECOCIOUS MENSTRUATION.

After the foregoing pages were in type, our attention was attracted to an article in the Cincinnati *Jour. of Med.*, Aug. 1866, from the pen of Dr. T. Parvin, of Indianapolis, Indiana, most of which we copy, although some of the cases he mentions are already included in our compilation or table on pages 76–78, *ante*.

" In *Meckel's Archives*, 1827, an example of early menstruation is given. At birth the child was of an ordinary size; but after the first month she commenced to grow rapidly, and at nine months was of the usual size of a child of a year and a half old. About this time she passed from the vagina a few drops of blood; at eleven months of age she had another and more abundant sanguineous discharge; and at the same time the mammary gland began to be developed, and hair appeared on the mons veneris. At fourteen months a third, and at eighteen months, a fourth sanguineous evacuation from the vagina. The whole physical development of the child was precocious; but her mental faculties were not greater than those of other children at her age.

" Dr. D. Rowlett, of Kentucky (*Trans. Jour. of Med.*, Oct. 1834), records the case of Sally Deweese, born April, 1823; menstruated regularly from the time she was twelve months old, until 1833, when she became pregnant, and in April, 1834, was delivered of a healthy female child, weighing seven pounds and three quarters.

"The third case is that recorded by Dr. Carus, of Dresden, 1842, in these words: Christiana Theresa, a natural child, was born in the mountains of Saxony; neither her father nor mother was of a robust constitution. She was scarcely a year old when she began to grow rapidly. At the end of her second twelve months the catamenia appeared, and have continued ever since to flow regularly once a month. The Academy of Medicine, of Dresden, sent for both her and her mother, and in order to examine more particularly into the case, kept them under their observation during several weeks. The infant was thirty-seven inches three lines long. The mammæ were firm, like those of a girl of sixteen. Her body was stoutly made, and the genital organs were covered with dark-brown hair. The physiognomy, and tone of voice were childish, which contrasted singularly with the strength of her body. Her intellectual functions were equal to those of a child three years old, and her head was covered with beautiful dark-brown hair.

"Mr. Whitman (*Northern Journal of Medicine*, 1845), relates the case of a child, who from a few days after birth, had her catamenia regularly at periods of three weeks and two or three days, until her death, at the age of four years and some months. At this last period the development of her body was very striking, equaling that of a girl ten or eleven years of age. The mammæ were unusually large;

the mons veneris collapsed, but well covered with hair—the labia pudendi sparingly so, though these organs themselves were of an unusual size for a child. The development of the pelvis and of all the deep seated genital organs was very considerable; and the lower limbs were proportionately large.

" In the *London Medical Gazette*, Nov., 1848, there is recorded the case of a girl who commenced to menstruate at the age of ten years and six weeks, and had a regular return of the catamenia in somewhat profuse quantity until conception, and at the age of twelve and a half years, she gave birth to a living and, for the most part, healthy infant.

" In an abstract of Dr. Szukit's investigations in reference to *Menstruation in Austria* (*Hay's Journal*, Oct., 1858), the following is found: Although in Austria, menstruation most commonly occurs between the fifteenth and seventeenth years, many cases have been recorded of its much earlier occurrence. Wilson observed a case in which it occurred in the fifth year; the breasts being as large as those of a marriageable female. Scanzoni records one case at eight years; D'Outrepont, one at nine months; she had protuberant breasts, and menstruated every four weeks till her death in her twelfth year. Boisment records two cases—in one, menstruation commenced in the third month; in the other, in the third year. Similar observations are recorded by Madame Boivin, Dieffenbach, and Martin Wall. In France early menstruation is more common than in Austria, as Boismont in 1,200 cases found fourteen who menstruated before their tenth year.

" In Cazeaux's *Midwifery* (3d *Am. Ed.*, p. 84), we find mentioned on the authority of Dr. Susewind, the case of a child seventeen months old, who menstruated regularly, and had, since she was twelve months old; her breasts and mons veneris, were those of a girl of fourteen or fifteen years of age; and on that of Lenhossek, another case in which menstruation commenced at nine months, and at two years the subject presented all the external signs of puberty.

" We will now present a few particulars in reference to a case which we have recently seen at Shelbyville, Indiana, of a girl four and a half years old. Her mother is under medium size, very delicate; menstruated first at twenty years of age; married at twenty-four, and two years after marriage, gave birth to this her only child. This child commenced to menstruate at three years and six months of age. Her general appearance is that of a stout healthy girl of ten or twelve years; her weight seventy-five pounds; her height three feet eleven inches; her voice rather coarse and harsh, at any rate it has not the softness and gentleness of infancy; her physiognomy that of early childhood; she is timid and " babyish ;" mentally and morally, she presents none other than the characters which might be expected in one of her age. But her most marked physical characteristics are those of a sexual sort. The mons veneris, though destitute of hair, and the labia, are well developed, and the mammary glands are quite large and well formed; in size they might answer very well for one sixteen or eighteen years of age. The circumference of the chest, measuring over the mammary glands, is twenty-seven inches; a line encircling the lower part of the trunk and fixed at either side at the middle of the crest of the ilium, measures thirty-one inches. The menstruation recurs regularly, and continues three days; she does not seem to have any special suffering at these times; the amount of catamenial discharge is about equal to the average observed in the adult during the same length of time."

SUPPRESSED MENSTRUATION.

Dr. N. D. Tirrell, in the St. Louis *Hom. Obs.*, of February 15, 1867, reports the case of a young woman, aged twenty-eight, married at twenty, the mother of three healthy children. Between the weaning of the first child and the birth of the second, was an interval of fourteen months, and the same interval between the weaning of the second child and the birth of the third. In each interval there were periods of five months elapsed between the weaning of one child and the conception of the following. The unusual and remarkable, if not extraordinary, fact is, that this woman had no return of the menstrual discharge during these several intervals, nor at any time from three weeks after marriage until December, 1866—a period of about seven years; neither had she any "vicarious" menstruation or discharge of any description whatsoever, except that soon after weaning her first child, she had what she thought was a miscarriage after three months' conception. She had no physician at the time, and, as far as she knows, there was no delivery of a fœtus. Dr. T., by a course of medicine, produced two returns of the long-suppressed monthly flow, but was not satisfied that it was regularly re-established.

INFLUENCE OF DRUNKENNESS ON CONCEPTION.

Dr. Demeaux adduces strong facts in support of the proposition that conception during drunkenness is one of the causes of epilepsy, and of other affections of the nerve-centers. He attributes to the same cause a great number of monstrosities and malformations, congenital lesions of the nervous centers, etc., which prevent complete evolution of the offspring; or if it arrive at term, cause early death.

MENSTRUATION IN RELATION TO PREGNANCY.

Conception, says Prof. Dubois, may take place in a woman not yet arrived at the age at which she ought to menstruate. There are, in fact, many women who do not menstruate till the 17th, 18th, or 19th year. Such women may become pregnant at that time of life, although they have never menstruated. M. Dubois has known a woman become pregnant two years after the cessation of the menses, and in due time labor supervened.

"The menses may be suppressed physiologically, and yet pregnancy take place. Thus, it is not rare to see nursing mothers become pregnant before menstruation has reäppeared. Women who are extremely irregular—who, for example, menstruate only once or twice a year—become pregnant; although, in general, this state is one very unfavorable for conception.

"Various diseases and changes in habits may derange the menstruation, and give rise to the idea of pregnancy. This error occurs frequently, chiefly to persons anxious to become in the family way. Nothing is more common than to find the menses suppressed for some time after marriage. It is also very frequently observed that women leaving the country to reside in the town suffer from suppression. This may be said to occur habitually in young women coming from the country into domestic service in Paris.

"In other cases, the menses, after having been suppressed for three or four months, reäppear suddenly with some profuseness. This is sometimes taken for the appearance of abortion, when it is merely the recurrence of the menses after they have been suppressed, in consequence of some change in the habits of the female."

,

MENSTRUATION DURING PREGNANCY.

A discharge, more or less identical with the ordinary catamenial flow, *may* occur during pregnancy. Dr. G. Hewitt records the case of Mrs. M. B., aged twenty-five, married six years, the mother of two children; during each pregnancy, every fourteen days a bloody discharge occurred, lasting three or four days; and this periodic discharge persisted during the whole period of gestation. During lactation, or the suckling of her child, no trace of bloody discharge was noticed.

DURATION OF PREGNANCY.

Some women experience peculiar sensations at the time of conception; hence, these women, reckoning from that date to the period of their delivery, have usually found the duration of pregnancy 280 days. In some cases a solitary intercourse fixes beyond question the date of conception. The lapse of 280 days thence to delivery, in a number of such cases, has made that the most usual duration of pregnancy. Some careful inquirers, upon the ground that comparatively few women are conscious of the moment of conception, give, as a rule to ascertain the most probable day of a woman's confinement—add 278 days to the last menstruation.

Dr. Murphy, of London, kept a record of 965 cases, which he divided into four tables: 1st, 303

15

cases, which exceeded 280 days of duration of pregnancy; 2d, 378 cases, which were exactly that period; 3d, 201 cases, ranging between 260 and 280 days; 4th, those below 260 days, embracing 83 cases. The last cases he looks upon, and very properly, as instances of premature labor. From these researches, he gathers that the duration of pregnancy is a variable period—260 days, or 37 weeks, being the shortest period; he has attended mature infants born at that period. He concludes his report of same, October, 1851, with a case of protracted pregnancy, six and a half months after quickening, or two months longer than usual.

Dr. C. Joynt, in the Dublin *Jour. Med. Sci.*, November, 1866, reports a case of a lady thirty years old, whose seventh pregnancy lasted at least 317 days, or six weeks more than the average. She suffered from excessive menstruation and neuralgia of the ovaries, and was also subject to frequent hystero-epileptic fits. The case was most carefully investigated, because of its extraordinary character.

The Code Civil of France provides that a child conceived during the marriage has the husband for its father; nevertheless, he may disavow the child, if he can prove that during the time that has elapsed between the 300th and the 180th day before its birth, he was prevented, either by absence, or in consequence of some accident, or on account of some physical impossibility, from cohabiting with his wife. The legitimacy of a child born 300 days after the dissolution of the marriage may be contested.

The Civil Code of Louisiana provides that the child born alive before the 180th day after the marriage, and also the child born 300 days after the dissolution of marriage, whether by divorce or death, is not presumed to be the child of the husband.

The statute law of Kentucky provides that any person born within ten months after the death of the intestate, shall inherit from him as if he were in being at the time of such death.

DURATION OF LABOR.

We have already mentioned, on page 212, the observations of Dr. M. C. Richardson, of Hollawell, Maine, as to the duration of labor in 1,000 obstetrical cases in his practice. As accurately as he could ascertain it, from the patients themselves, the average duration was eleven hours and thirteen minutes; the longest labor, with regular pains, ninety-six hours; the shortest, one single pain. This last almost equals the cases of painless and unconscious parturition, on pages 183–8.

In 311 cases, recorded by Prof. Simpson, in the Maternity Hospital of Edinburgh, Scotland, the average duration of each case of labor was thirteen hours.

Dr. James N. Fraser, a Scotch physician, at St.

John's, Newfoundland, reports the average duration in 93 cases of labor in his practice at thirteen and one-half hours.

The average of the above 1,404 cases would be eleven hours and three-quarters each.

SUSPENDED FŒTAL LIFE.

In a lecture in 1860, of Dr. T. Gaillard Thomas, of Bellevue Hospital, New York City, on "The Accidents which may occur subsequent to Parturition," it was stated that the tenacity of the fœtus is remarkable, and the neonatus will survive without air much longer than its parent. Many a new-born child, to all appearances perfectly dead, has been resuscitated—in some cases even after the heart's action had entirely and certainly ceased; in one case, in Dr. T's hands, after three-quarters of an hour constant exertion, and when twenty-five minutes elapsed between the delivery and the appearance of the first visible sign of returning animation.

In 1850, in France, a woman, with the intent of infanticide, buried her child just after delivery, in the earth. After forty-five minutes it was discovered and disinterred. It was found lying on the placenta, which was still attached, and restored to life.

A case is reported in the Cincinnati *Lancet*, where the woman died in labor before the physician, Dr. Thornton, could reach her. Dr. T., delivered the child forty-five minutes after the mother's death—

itself still and apparently dead, but after half an hour's artificial respiration it was resuscitated.

Dr. Jenkins, of Yonkers, New York, delivered a child, the funis of which had prolapsed into the vagina and been pulseless for twenty-five minutes before birth. After birth, for half an hour, no effort at resuscitation was made; then *two hours* of active exertion succeeded, and the child was restored to life.

Several other cases of the same sort are recorded —proving that even where the heart is motionless, and the child to all appearance entirely and certainly dead, vigorous efforts to resuscitate have been successful at all periods of time afterward, from three minutes to an hour and a half.

SUPPOSED SUPERFŒTATION.

Dr. Atchison communicated to the Obstetrical Society of Edinburgh, the case of a well-formed woman, aged thirty, who was prematurely delivered of a seven month's child, which survived only a few hours. To the placenta he found attached a large sac, which proved to contain another fœtus, supposed to have completed nearly its fourth month. He mentioned a number of circumstances which made him regard the case as one where one ovum was impregnated previous to the other, and, therefore, of true superfœtation.

Dr. Keiller thought the case one of twins, where

one of the children had died. Many similar cases had been recorded. It often happened that one child died in utero and was retained, while the other went on to the full time. When the liquor amnii is retained, the dead child may be kept in perfect preservation for a very long time. In his museum he had a number of preparations illustrative of blighted twin cases. The blighted fœtus usually presents a wizened or squashed appearance. A case occurred in the Maternity Hospital, where the blighted fœtus was not discovered for a day or two after the delivery of the other child. On examining the placenta, a small clotted mass was observed attached, and, on raising the membranes, the blighted twin was found. Most of the so-called cases of superfœtation were of this nature. In true superfœtation both children are at the full time, and the one is expelled a long time before the other. Such cases are very rare, but they may occur. Dr. Matthews Duncan shows that the spermatozoa may reach the ovule for three months after impregnation has taken place. The decidua does not obstruct the passage of spermatozoa in the early months. The old notion was that the mucous plug prevented their passage. Another explanation of superfœtation is the existence of a double uterus. The chief difficulty in the matter, however, is that during pregnancy there are no ovules to impregnate. Scanzoni at one time held that ovulation went on after impregnation had taken place, but he has now changed his opinion.

Dr. Cuthbert had met with a case similar to that related by Dr. Atchison, in which one of the children had died and the other was born healthy. The dead fœtus was withered and shriveled.

A FŒTUS CARRIED FOR FIFTY-FOUR YEARS.

The Edinburgh *Med. Jour.*, September, 1862, contains a well authenticated account of a post-mortem examination of a woman, from whom was extracted (extra uterus) a tumor or bony cyst, containing a fœtus which must have lived a long time after the natural term of birth, and with which the mother was pregnant fifty-four years; having given birth to two living children at four and eleven years respectively, after such extra-uterine pregnancy began and during its continuance.

UTERINE CRYING.

The possibility of a child in the womb having its lungs so filled with air as to be able to utter an audible cry, has been doubted. But several cases are reported in medical journals which remove all doubt on the subject. In all authentic instances the cry has been heard only where the face has presented, and the accoucheur has by any accident introduced his fingers into the mouth of the child, and thus made a passage for air to the lungs.

INFANTILE SYPHILIS.

Dr. R. Förster, of Dresden, Saxony, gives the results of his observations, during nine and a half
years, on sixty-eight cases of infantile syphilis.
Appearing in newly-born children or soon after
birth, it is almost inevitably fatal; with more hope
of cure the later in infantile life it appears. The
sixty-eight children (twenty-eight males and forty
females) varied in age from twelve days to four and
a half years; 45 (about 66 per cent.) recovered; and
23 (or 34 per cent.) died. Of 36 children who continued to suckle their own mothers, during medical
treatment, only 6 died; while of 18 children deprived
of the breast, 13 died. The duration of treatment,
by Dr. F., varied from two and a half to thirteen
weeks; the average being five and three-quarter
weeks.

INVERSION OF THE HAIRS OF THE LABIA PUDENDA.

In 1862, Dr. Chas. D. Meigs recorded the case of
a young lady, aged twenty, who suffered greatly
from continual pruritus and heat of vulva. After
other physicians and many remedies had failed, he
discovered that the margins of each of the labia were
studded with long, straight, and stiff hairs, just like
eyelashes—all directed inward, and so, constantly
teasing, irritating and vexing the mucous body of
the interior, and producing a redness or florid tint of

the membrane, with heat and the intolerable itching of which she so long complained. They were gradually eradicated by tweezers, by the lady's nurse, after some days; and the lady thus cured of what had seemed an incurable disorder.

ABSENCE OF UTERUS AND OVARIES.

Dr. S. Hertz, of Boonville, Indiana, reports, in July, 1870, the case of Miss E., aged forty, unmarried, who had enjoyed good health until during the last year of her life. About three months before her death she came under the care of Dr. H., for a slight dyspeptic disorder, which, on close examination, proved to be due to compression of the stomach, consequent upon an enormously enlarged cancerous liver.

The *post-mortem* examination, made by Drs. Barker, Darby, and Hertz, showed the liver to be greatly enlarged by cancerous deposit; stomach and intestines free from any deposit; kidneys somewhat enlarged and indurated. The chief point of interest, however, was a complete absence of the uterus and ovaries. The vagina was normal, both as regards length and capacity, terminating above in a cul-de-sac. The clitoris was well developed, together with the labia and mons veneris. The breasts were large and plump, the whole external aspect attested the attributes of a well formed woman.

CHAPTER XV.

PATERNITY.

IN the law case of Plawes *vs.* Bossey, in England, in February, 1862, the legitimacy of a person was disputed on the ground that the father, at the time conception must have occurred, was confined in a lunatic asylum, and had no access to his wife. The rules of the asylum, when the wife visited her husband, forbid their being together under such circumstances of privacy, etc., as would permit of connection.   But it was shown that the lunatic was frequently alone and unobserved for two hours, and had been to the neighboring town.   The wife claimed that the private interview took place at a friend's house in that town.

The Vice-Chancellor gave judgment in favor of the legitimacy—observing that a child born of a married woman was always presumed to be legitimate, that is, the child of the husband.   The *onus* was on those claiming the contrary; they *must* prove *im*possibility of access.   No *onus* lay on the other side to prove possibility of access.

TRIALS FOR RAPE.

In a case tried in the Supreme Court of New York, in 1867, of Walter *vs.* The People, it appeared that a woman of weak mind, but not imbecile, thirty years old, had been for some time under the care of a physician for a temporary complaint. When cured of this, she informed him of another complaint, which he pronounced womb disease, and said he would have to make an examination of her person to inform himself of the precise nature of her disease. She at first objected, but, upon his insisting that it was absolutely necessary and could not be avoided, she consented. While professing to make this and a subsequent examination, he had connection with her twice—in the parlor of her brother's house, in the day-time, while her brother's wife was in an adjoining room. She made no outcry, believing that what he did was what he declared it to be, only a medical examination in the usual way. She did not know better, and did not tell of the circumstance until after she was charged with pregnancy. The physician was arrested, tried for rape, and sent to the penitentiary—whence he was released by order of the Supreme Court, which decided that although her seduction was accomplished by fraud, it was without the force or violence, or threat of either, necessary to constitute it a case of rape. The case is reported in the American *Law Register*, Oct., 1867. The judges drew a distinction between actual assent to sexual intercourse, and acquiescence by passive non-resistance in an act of the physician, of the nature of which she was ignorant. They evidently believed that there was such consent or willingness as took away the offense charged.

In the case in England of Regina *v.* Case (1 *Eng. L. and Eq.*, 544), a medical practitioner had sexual connection with a female patient of the age of fourteen years, who had for some time been receiving medical treatment from him for illness arising from suppressed menstruation. The jury found that she was ignorant of the nature of the physician's act, and made no resistance, solely from a *bona fide* belief that he was (as he represented to her) treating her medically, with a view to her cure.

In another case in England, Regina *v.* Clark (29 *Eng. L. and Eq.*, 542), the prisoner got into the bed of a married woman, intending to have connection with her by passing for her husband, but not by force. She, supposing him to be her husband, submitted to his embraces. The court held that the prisoner was not guilty of rape.

INSANITY FROM MASTURBATION.

Dr. Choate, of the Insane Hospital at Taunton, Massachusetts, ascribes to acts and habits over which the individual has full control, 421 out of 1195 cases in which the cause of insanity was known. It is probable there were many more cases, where, from shame and dread of disgrace, the true cause was carefully concealed. Habitual stimulation of this kind is the source, Dr. C. thinks, either directly or indirectly, of more mental disease than all other causes of insanity combined. "Acting upon the body, it is the immediate cause of many cases of mania and dementia, of nearly all the instances of softening of the brain, of epilepsy, paralysis, and diseases of the digestive organs which, in their turn, rëact upon and overthrow the mind. Acting upon the mind, its tendency is to weaken the reasoning faculties, and undermine the judgment, inducing unwise business transactions, unfortunate social connections, loss of property, and the mental disorders which these occasion. Acting upon the heart, it perverts the moral sense, and weakens the powers of resistance to temptation, leading to domestic unhappiness, ill-treatment of friends, and indulgence in other fatal habits; and these, in turn, contribute their share to the numbers of the insane."

UNITY AND NON-UNITY OF THE NEGRO AND WHITE RACES.

In a letter to the Richmond and Louisville *Medical Journal*, September, 1868, Dr. E. B. Turnipseed says: "I am not aware that it is known to the scientific world, that the hymen of the negro woman is not at the entrance of the vagina, as in the white woman, but from one and a half to two inches in the interior, with a passage below for the escape of the menses. I have examined a good many cases, and have found this invariable. This I thought to be abnormal at first; but finding it constantly situated as above described, in examining cases during a practice of fifteen years, I have concluded that this may be one of the anatomical marks of non-unity of the races. I will say, further, that in this race I have never found the hymen situated as in the white race—at the entrance of the vagina."

WEANING.

Some young mothers stop nursing at the earliest moment their doctor will sanction it, while a much larger class, for various reasons, wish to nurse as long as they can. The collated opinions of many authors and of many active medical observers agree upon nine months as the most reasonable and most approved term of lactation. Prolonged nursing is injurious to both child and mother—causing in the child a tendency to brain disease, and in the mother to deafness and blindness. Teeth-cutting is really of very little consequence in deciding *when* a child should be weaned.

As to the best *season* of the year, the testimony of both authors and practical men is strong and decisive against weaning in hot weather; and points clearly to the late *fall* as the best period. But if, early in May, the child be nine months old, wean it, unless its mother exhibits uncommon capacity as a nurse. Each case, however, must be considered and arranged by itself.

LACTATION.

In an aged Female.—Dr. Gillespie, in the Boston *Med. and Surg. Jour.*, Aug., 1868, relates that Mrs. Arnett, a widow from Tennessee, aged sixty, came to Virginia with her youngest daughter, aged thirty—who married there; in time was delivered of a small feeble child, and died when it was about two months old. The grandmother, a robust, healthy woman for her age, being unable to procure a wet-nurse for the babe, was persuaded to try the child to her own breasts. She did so perseveringly, and a plentiful supply of milk was secreted. She continued to nurse the child, which became strong and healthy, until it was nearly two months old.

In a Male.—A young man, aged twenty-two, finding his left breast enlarging and somewhat painful, went to Philadelphia, to the faculty of Jefferson Medical College. Prof. Mûtter found the mammary gland largely developed, and filled with lacteal secretion, which differed in nowise from that of a mother. Six weeks' treatment reduced the breast to the usual size.

INFANTS WITH TEETH AT BIRTH.

In a note to the *Am. Med. Times*, Dr. Elliot gives the following cases : " In the lying-in wards of Bellevue there are now two infants under my care who were born with teeth, viz.; William Hoffman, sixth child, weight at birth six and one-half pounds, puny, fully developed; right middle incisor in lower jaw well formed and protruded, but placed athwart the jaw. Annie Morse, first child, weight at birth seven pounds, fully developed; two middle incisors in the lower jaw, both well formed, but loose; right incisor set obliquely in the alveolar process."

ROSE-COLORED TEETH.

Professor Moritz Heider, of Vienna, relates that two girls (twins), who were placed under his care, had teeth of a peculiar rose-color. On the shedding of the first teeth, the permanent set also appeared of the same rose-red color, and only paled off after some years, never losing the reddish tint entirely. This appearance is difficult to account for, as no other members of the family shared the same peculiarity, nor was there any difference in the mode in which they were brought up.

EARLY PARENTAGE.

On the 18th of June, 1869, in the city of New York, as recorded by the *Express*, a young lady, aged fourteen, living in a very aristocratic region in the city, was taken suddenly ill, and appeared in such extreme pain that the family became alarmed and sent for a physician. He came in all haste, and, after glancing at his patient, requested the parents to leave him alone with her; they did so, and, in half an hour, another young lady opened her eyes for the first time on this world of sorrow. She made such a noise coming into this world that the parents' attention was attracted, and the doctor informed them that they had better send for a nurse, as mother and child were " doing as well as could be expected." Great consternation reigned. On being closely questioned, the young mother placed the paternity of the stranger upon a lad of fifteen, who was then at school. Charley was sent for, and blubbered like a baby. There was then in the room a father, a mother, and a child, whose united ages amounted to only *twenty-nine*.

A still more remarkable case of early parentage was that (already mentioned on page 80) of Elizabeth Drayton, of Taunton, Massachusetts, who, when only ten years, eight months, and seven days old, was delivered of a healthy male child, at the full time of pregnancy. The reputed father of the child was a lad from Maine, about fifteen years old. The amorous young couple were detected *flagrante delictu*, and the boy banished to his home as a punishment. No one dreamed of any serious result; but on Feb. 1, 1858, just nine months after, as appears from well authenticated records, a nice, full-grown, plump baby, weighing eight pounds, good weight, was introduced into the world—who, in November afterward, was des-

16

cribed as a fine little fellow, of a very handsome model, hair curling a little, a bright blue eye, healthily grown, and with all essential elements in him to make a great man. These parents and child were unitedly less than twenty-six years in age.

Mrs. Eunice Warner, formerly of Great Barrington, Massachusetts, became a mother at thirteen years, a grandmother at twenty-seven, a great-grandmother at forty, a great-great-grandmother at fifty-six, and a great-great-great-grandmother at seventy-four, after which she lived several years.

LATE MATERNITY—LARGE CHILDREN.

In January, 1869, old Johnny Grim, of Newville, Johnson County, Indiana, aged seventy-six, was presented with a bouncing boy by his kind-hearted " old woman," who was over sixty years of age.

On the 26th of May, 1870, a lady in Berkshire County, Massachusetts, already the mother of ten daughters in succession, was delivered of her first son. In a little over two months afterward, she entered on her forty-ninth year. We learn this by a letter from the husband of the lady.

Dr. Gruwell, of Iowa, records that in August, 1868, he assisted in a difficult obstetrical case, where Mrs. Y., aged forty-eight, and mother of ten children, was delivered of a dead child, weighing eighteen pounds.

The wife of Milton Perry, near Skinnersville,

Scott County, Kentucky, on August 12th, 1870, gave birth to a living child weighing sixteen pounds.

BABY SHOWS.

Quite a sensation was created all over the country, some years before the war, by P. T. Barnum's idea of a set of baby shows, systematically and successively held in the large cities of the United States. They were attended by immense crowds of curiosity-loving people, and "put money in the purse" of the irrepressible showman.

Profiting by his experience, a similar show, under the management of C. St. John, was inaugurated at Boston, Massachusetts, in June, 1869. One hundred and seventy-two babies were exhibited, each one receiving a present of a silver cup at the close. In addition, handsome premiums in money were awarded to the following as the *prize babies :*

"*Triplets*—Nellie Hope, Horah Faith, and Hannah Charity Coughlin, of North Brookfield, six months old, the only premium for this class.

"*Twins*—James L. and George W. Duval, five years old, sons of George W. Duval, of Chelsea, first premium.

"*Handsomest Girl*—Allie E. Stratton, twenty months, daughter of Alfred A. Stratton, of Boston (who lost both arms in battle in June, 1864), the first premium.

"*Handsomest Boy*—Clarence McGuire, three years old, son of Rebecca McGuire, Boston, first premium.

"*Fattest Baby*—R. N. Ash, Sommerville, nine months, first premium.

"*Smallest Baby*—Alice Hillock, Charlestown, two months and

five days old, weighing four pounds, and only one pound when born, first premium."

It was this baby show, or something else of a sensational character about the same time, which brought out the following article from a correspondent of the ladies' newspaper at New York, *The Revolution:*

"WARNING TO HUSBANDS.—The great want of women at present is money—money for their personal wants, and money to carry out their plans. I propose that they shall earn, that they shall consider it as honorable to work for money as for board, and I demand for them equal pay for equal work. I demand that the bearing and rearing of children, the most exacting of employments, and involving the most terrible risks, shall be the best paid work in the world, and that husbands shall treat their wives with at least as much consideration, and acknowledge them entitled to as much money, as wet nurses. The meaning of this is, that wives are about to strike for greenbacks; so much for every baby borne. No greenbacks, no more sons and daughters. No greenbacks, no more population; no more boys to carry on the great enterprises of the age. The scale of prices for maternal duties are given as follows:

Girl babies	$100	Boy babies	$200
Twin "	300	Twins (both boys)	400
Triplets	600	Triplets (all boys)	1,000

Terms: C. O. D. No credit beyond first child, the motto being, 'Pay up or dry up.' Husbands who desire to transmit their names to posterity will please notice and take a new departure."

PROLIFIC ANIMALS—TWINS.

A mare belonging to Martin Lydick, who lived between Leesburg, Harrison County, and Newtown,

Scott County, Kentucky, gave foal to a colt on Sunday, January 27, 1868, and another on the Tuesday following, and the very next Thursday added a third to the number. She was reported as defying any matron of the horse kind to beat that.

Elijah Pollard, in Henry County, Kentucky, put up a sow, a few days before the heavy sleet in April, 1868, to have pigs. She brought forth four or five, which died in a few days from the effect of the bad weather. Mr. Pollard then turned the sow out, and in less than three weeks she bore three other pigs, which, two months later, were living, squeaking pigs.

On the 12th of July, 1870; a sow belonging to Robert M. Lee, of Scott County, Kentucky, had a litter of three pigs, and seven more on the 19th of the same month.

A correspondent of the London *Gardeners' Chronicle*, March 21, 1863, says he has a ewe that dropped a ewe lamb on the 18th of February, and a male lamb a month after.

Another correspondent of the same paper, April 18, 1863, says a ewe of his flock dropped a dead lamb about the 25th of January previous, and was then turned over to the part of the flock that had lambed. Her milk did not come then, and about the 24th of February she had another lamb, still alive when he wrote. He had never seen any other case like it.

About the 15th of June, 1870, a cow belonging to Wm. Burnell, of Swanton, Vt., gave birth to a calf, which in a few days was killed, and the milk of the cow saved. The milk was very thin and appeared to have no more richness than ordinary skimmed milk. And yet the cow appeared well, and the cause was quite a mystery. A month afterward the cow had another calf, and since then her milk has been perfectly good.

A BREEDING MULE.

A female mule at Mont-de-Marsan, France, of twelve years of age, has dropped a mule colt, born at term and perfectly formed. The dam gives milk and the foal sucks, but the mother manifests a profound indifference for her offspring, and does not exhibit the slightest solicitude when separated from it.

CHAPTER XVI.

PREVENTION OF OFFSPRING.

WE have already alluded, in Chapters I and II, to the love of offspring as the ruling passion in woman's nature. Perhaps this should be qualified by saying it is part of the nature of every *true* woman. And yet there are many noble, gentle, loving wives, who—because of the bodily suffering incident to childbearing, because of its heavy draft upon the physical strength and the constant strain upon the nervous system, and because of other reasons not less satisfactory and soothing, and consistent with the wishes of the sufferer—most beseechingly and persistently implore their husbands to spare them the painful infliction, and " wait a little longer." This growing aversion—growing even among good and conscientious women—to having children frequently, is telling with powerful effect upon the public morals. If betrayed by the exciting or amorous feelings of the moment beyond the bounds of self-denial or cautious self-control, they suspect that the fatal step has been taken, and the hated seed planted in a sure place— not all the love they bear their husbands, nor the

simple-hearted and sincere regard for duty which erst
has kept them in the right path, and sometimes not
even the humble fear and reverence with which they
seek to know and obey the will of their Heavenly
Father, will reconcile them to the sore disappointment, and make them submissively and hopefully
look forward to the hour when they shall forget their
pains in the joy that a "man is born into the
world." It is not alone those who through "not
wisely loving, but too well," or through the sin that
has not even the semblance of innocence, but those
also in the position of wives, and with the prospect
or promise of mothers, who yet are unwilling to
become mothers—that rebel against the fate which
overtakes all who yield to the caressing embrace ; and
whose rebellion too often leads to crime—crime
against themselves, crime against the unoffending
beings that have not seen the light of heaven, crime
against that society which has tolerated and protected
them, and crime against the God that created them
and the helpless innocents within them.

We call it crime to produce abortion or miscarriage—crime to shut out from life that in which life
has begun. No softer word will express the act,
and very seldom will any state of case excuse or justify it. Most heartily, therefore, do we approve and
uphold the action of the Academy of Medicine of
Cincinnati—who, in regular session, May 25, 1868,
unanimously adopted the following resolutions, submitted by a special committee composed of Drs. J.

F. White, James Graham, F. G. Schmidt, C. S. Muscroft, John Davis, and John L. Neilson, appointed to report upon "procured abortions and criminal advertisements : "

" 1. Criminal abortions are fearfully frequent.

" 2. That, as a general rule, the crime is committed by irregular practitioners of medicine, by certain female *accoucheurs*, and by apothecaries who vend certain nostrums to correct suppressed menstruation.·

" 3. That we believe that the advertisement of abortionists and abortion drugs encourages the practice of abortion, and is criminal, and, therefore, the Academy of Medicine should earnestly protest against the admission by the press of such advertisements.

" 4. That the Academy appoint a standing committee to fortify, as much as possible, the Health Officer in Cincinnati in the prosecution of such offenders.

" 5. That it is the duty of all good citizens, and especially physicians, to discourage the circulation and patronage of the journals in which are published the advertisements of those who profess to produce abortion or prevent impregnation."

Still stronger and more condemnatory, and treading boldly upon higher ground, where many physicians refuse or fear to tread, was the action, in October, 1867, of the Clarke County (Indiana) Medical Society, in adopting the following :

" *Resolved*, 1. That it is the sense of this Society that any attempt to destroy the ovum or fœtus *in utero*, at any time, from the hour of conception to the end of *utero gestation*, or the maturity of fœtal life, is infanticide in the strict meaning of the term, and should incur all the penalty which is attached to murder, and the act should be punished as such, when brought to light.

" 2. That every physician who lends himself, in any manner, to this crime is *particeps criminis*, and should be punished accordingly."

But while the good and brave of the medical profession do not hesitate thus to put themselves upon record, the same class of men are just as decided in their convictions that many circumstances arise where the duty is clear to *induce* premature delivery—in such case no longer called abortion. Prof. T. Gaillard Thomas, of New York, in a valuable paper on this subject, in the *N. Y. Med. Jour.*, February, 1870, thinks that the most serious complications of labor, both as regards mother and child, may, in most instances, be recognized by their peculiar premonitory signs, one, two, or even three months before the end of pregnancy ; and being thus recognized, may often be avoided. He advances the opinion that nothing will in the future tend to diminish the mortality attendant upon parturition so markedly as the induction of premature delivery— for the rescue of the mother and child, or both, from the danger which threatens them. The morbid states for which he considers the operation advisable, are the following :

1. Deformity of the pelvis.
2. Placenta prævia.
3. Aggravated uræmia.
4. Excessive vomiting.
5. Placental apnœa.
6. Commencing epithelioma.
7. Death of child and consequent septicæmia.
8. Threatened death of child.
9. Approaching death of mother.
10. Amniotic dropsy.

11. Previous rupture of uterus or performance of the Cæsarean section.

12. Excessive accidental hemorrhage.

13. Previous difficulty in deliveries of large children, or of children with ossified sutures.

14. Tumors obstructing the pelvis.

Each of these conditions Dr. Thomas discusses in turn, giving clinical cases to illustrate the propriety or necessity of resort to the operation. The end of the eighth month, *i. e.*, the ninth menstrual epoch, he regards the most favorable time for the induction of premature labor. The distinction he makes between premature labor and abortion is quite marked. " The former denotes a premature expulsion of the contents of the uterus ; the latter, a failure in the results of utero-gestation. Consequently, the induction of premature labor is in one essential respect different from that of abortion, and is called for in the fulfillment of different indications. The former, being resorted to after the period of viability of the child, does not involve the sacrifice of its life—but often adds to its prospect of living by removal of it from a position of danger, and sometimes even of certain death." Or, to put the matter more tersely, abortion is resorted to in the interest of the mother alone, at the expense of the life of the child ; whereas, premature labor is induced sometimes for the sake of the mother, sometimes for the sake of the child, and sometimes in the interest of both. His subject was the induction of premature delivery as a prophylac-

tic resource in midwifery; and in no way concerns
the induction of abortion.

If, then, in the judgment of some of the wisest and
most prudent in the profession, there are so many
circumstances arising which indicate that premature
delivery should be promptly resorted to—in honest
efforts to save the life of the mother and child both,
or of either if impracticable to accomplish both; it
would be singular if there were not many corresponding circumstances, in which it would be to the
same extent wise and prudent to forestall the necessity thus indicated—by employing some simple and
proper means to *avoid conception.* A little reflection
will remove all doubt as to the propriety of such a
course, provided any such safe and unsinful mode of
prevention is known and pointed out by the profession.

So constituted are some women that they can not
give birth to *healthy* children. Others again can not
bear *living* children. Others, still, have not health
and strength to nurse and nourish their children,
even after they are safely delivered; and after drag-
ging out a miserable existence, with added anxiety
and responsibility to the helpless parents, they lay
them away in an early grave. Other parents find
developed in themselves the seeds of consumption,
scrofula, and other inheritable diseases, which they
dread to hand down to a puny and afflicted offspring.
In one case within the knowledge of the writer, the
wife—remarkably well developed as a healthy woman,
and admirably adapted to raise children—mated with

a man more than twenty years her senior, of most
irregular and ungovernable temper, at times very
cruel and harsh, and otherwise of singular disposition and taste, was so impressed with the idea that
no offspring from such a parent ought to be raised,
that she practiced abortion upon herself, by her own
hand, for more than twelve years—justifying her conduct as simply *duty*.

These are but a few of many cases that would seem
to make it prudent to avoid offspring. Remember,
fourteen separate causes are mentioned above by
Dr. Thomas, as justifying or requiring premature
labor, to avoid greater evils. If so many, or more,
causes make it advisable to prevent the necessity of
such serious work—will they not make it much more
a humane course to forestall the necessity for it? If
this can be done, in a safe and proper manner, well;
if not, it would be better not to attempt it. In the
parable of the tares, in New Testament illustrations,
the servants of the householder asked whence came
the tares, knowing that good seed had been sown;
and when they proposed to go and gather up the tares
the householder refused, "lest while they gathered
up the tares they should root up also the wheat with
them." So, if the seed be already sown and have
taken root, in the seed-ground of families, it is far
better every way to let the impregnation go on to
child-birth, unless to save life or other great necessity indicate the duty of the physician to be to induce premature delivery.

The marriage service of the English Church states that the marriage institution is ordained for three purposes: "1. The procreation and the due education of children; 2. The avoidance of incontinence; 3. The mutual society, help, and comfort of the married pair." Any union of the sexes which does not provide for fulfilling every one of these purposes, or which fails in so doing, comes short of securing the great objects intended by this institution. Children must be procreated; they must be duly educated also. In order to this, they must be living, and must be healthy in body and mind. "Can a corrupt tree bring forth good fruit?" Can diseased parents bring forth healthy offspring? Incontinence must be avoided; not by crime—as abortion—but by reasonable indulgence, and by forbearance or abstinence when necessary to health. The married pair must enjoy each other's society, help, and comfort. How? Good health is at the foundation of all merely human enjoyment; good temper, gentleness, kindness, forbearance, are all incidents to good health, and indispensable to the right appreciation and use of the marriage relation. Nothing must be done, then, inconsistent with the preservation of health.

Not unconnected with this are the various causes which operate to obstruct the law of human increase. Fecundity in women, just as certainly as in the animal kingdom generally, is greatly limited, if not totally checked, by the plethoric state, while it is induced and increased by the lean state. Over-feeding—too much flesh—almost surely prevents reproduction. The owners of fine-blooded stallions are very cautious to avoid high feeding, because it hinders foal-getting. Lord Bacon very truly said, "Repletion is an enemy to generation."

High living, not in the sense just referred to, but in that implied by fashionable living, is an obstacle to propagation. Unsubstantial and highly-seasoned food, irregular hours of eating, insufficient rest, and the other incidents to a state of "high refinement," combine to lessen the number of children in a family, if not prevent them altogether. Besides, comparatively few children survive to manhood that mode of life.

It is a singular fact, when spoken of a large proportion of women in refined society in the United States, in the Eastern and Middle States especially—as remarked by Dr. Allen of New England women—that "the *sexual propensity itself* has undergone changes, not only in its activity and excitability, but particularly in its strength and powers of endurance. This propensity was planted in the human constitution by God himself for high and noble purposes. If a thorough discussion of this question could be had, based upon physiological laws, and connected with an exposure of the facts actually existing in the married state, it would furnish a key, not only to much of the discord in domestic life, but also to the increasing infidelity of husbands to their

wives, as well as to the great number of divorces
constantly taking place in the community."

EFFECT ON POPULATION.

Without pushing our suggestions in this direction
any further, is it any wonder that in view of these
facts, the population of several of the older of the
United States is positively decreasing, and in several
others the early population and their descendants are
slowly dying out?   In France, from the census, registration reports, and other official sources, it appears
that the population of that great nation has been increasing very slowly, while the number of births has
been decreasing steadily ever since the close of the
Napoleonic wars.   The average number of children
to each marriage in France, seventy years ago, was
five; but now, in the city of Paris, the average has
fallen to about two, while under the more favorable
influences of the country districts it does not exceed
three.   No nation, unless the recipient of an astound-
ing and unparalleled immigration, can increase in
population with such a birth-rate.   And hence we do
not doubt that the census of 1870 will exhibit not
only little or no increase in the black or colored
population, but over the whole United States a considerable actual decrease of population during the
last ten years.   A few months will tell.

In some practical investigations upon this point
Dr. N. Allen well observes that by census returns,

taken in 1765 and 1865—one hundred years apart—
there are now found only about one-half as many
children under fifteen years of age, relatively to the
adult population. In most parts of the New England
States there is no longer any increase of the strictly
native population. The native stock must hereafter
gradually diminish, while the total strength of population must be kept up and increased, if at all, by
the infusion of the more prolific foreign population.
The census of the State of New York, taken in 1865,
discloses some curious facts. Out of eight hundred
and forty-two thousand five hundred and sixty-two
women who then were or had been married, there were
115,252 who never had a child, or...... 13.69 per cent.
124,317 who had had only one child, or 14.75   "
123,319   "   "   " two children, or 14.63   "
108,324   "   "   " three   "   " 12.85   "

471,212, or nearly fifty-six (55.92) per cent. of all
the married women in the greatest of the States average only one child and seven-tenths to each mother.
Both the foreign and American classes are included
in these figures; and as the foreign class is largely
the most prolific, it follows that much the largest
proportion of the above are American families. If
we bear in mind the law of infant population, as
settled by the mortality tables, that two-fifths of all
children die before reaching adult life, it will appear
that these 471,212 mothers will hardly raise to manhood one child each. But, allowing that many of

17

these are young mothers, and will do better than this for their country, it is not probable that they would exceed the average of *two* children to each marriage—leaving all positive increase of population in New York as in New England, to come from the foreign population. In the county of New York, which includes the city, by the census of 1865, there were 965 American women who had each as many as ten children and upward, and 2,850 women of foreign birth with like numbers, or three times as many in about an equal population.

But enough as to the fact. A few words more as to the causes, in addition to what has already been said on pages 41 to 51, 169, 215, and elsewhere.

ABORTION IN INDIA.

While we specially avoid recapitulating here the names of the noisome and dangerous drugs which are accessible to the public in all large drug stores, and, because easily accessible, are extensively resorted to, we may do the honorable and judicious medical practitioner a service by stating that Dr. Shortt, of Madras, in Southern India, in a report to the London Obstetrical Society, in 1864, mentions the fact that twenty-seven persons were condemned in that presidency, in the previous year, for being concerned in criminal abortions—by employing the following among other substances (native to that country, and not accessible here) to expel the fœtus prematurely:

1. Bamboo leaves pounded in a mortar, the juice extremely powerful, causing expulsion of immature ovum in two to three days.

2. Juice of *Euphorbium Tirncali*, applied to the os by a rag on the point of a stick, insures expulsion in from twelve to twenty-four hours.

3. Other species of same genus accomplish the same.

4. So, also, the *Caloptris gigantea*.

5. So, also, the root of the *Plumbago Zeylanica*, used both locally and through the stomach; if used locally, the root is shaved to a taper point and so introduced into the os uteri.

SUGAR PREVENTING THE GENERATION OF ANIMALS.

Dr. Henry Tanner, Professor of Rural Economy in Queen's College, Birmingham, England, advanced the opinion, in 1866, that any animal may, by the too free use of sugar, be rendered incompetent for propagating its species. After his attention was first drawn to this fact, he observed numerous instances which tend to confirm this opinion. Among other instances, an eminent breeder, to improve the condition of his herd, added molasses to the dry food he gave his stock. It certainly improved their appearance and general condition; but another result followed, as unexpected as startling — his stock, always in demand at high prices for breeding, now, with very few exceptions, proved valueless for that object, male and female being alike sterile. Immediately upon this discovery, the use of molasses was discontinued. While the animals which had not been using molasses maintained their fine breeding qualities, it is very doubtful whether any that had been thus fed ever regained their breeding powers. It is not at all unlikely that a fatty degeneration of the ovaries took place, from which, under any ordinary treatment, their recovery would be slow.

In another case, where molasses had been used for some heifers which were fattening, it had the effect of suppressing those periodical returns of restlessness which prevent heifers feeding as well as steers; and it kept them steadily progressing during the whole period of their fattening.

If further observation and experiment should confirm this effect of sugar in checking the reproductive functions, while avoiding it for breeding animals, we may freely use it when cows or heifers have to be fattened.

The writer has for more than four years observed the effect, upon a young married couple, of the use of excessive quantities of sugar in their tea and coffee, and of unusual fondness for, and indulgence in, sweets of all kinds. Although descended from families embracing large numbers of children—their mothers respectively having borne fourteen and twelve children, and several of their brothers and sisters proven quite prolific—yet, so far as any indications are reliable, the wife seems permanently sterile, while the husband is probably likewise affected. There is nothing, apparently, in their physical conformations or temperaments to forbid a large progeny—which every thing would seem to favor except for this singular and noticeable devotion to the use of sugar, molasses, and other sweets. And why may not the human organs of generation be rendered unproductive by the same causes which are known to have affected animals?

But it may be well to give here a word of severe warning. There is nothing that would more certainly prevent marriage itself than the knowledge that a young man or young woman is permanently barren, or probably so. The existence of that fact, so soon as known, would break off every contract of marriage, hardly excepting one in ten thousand. The love of offspring is so much a part of human nature, that no misfortune is more humiliating to a sensitive nature than the fact of absolute inability to produce or bear children; and to have that fact *known*, we are confident, would drive the afflicted one to a change of residence more certainly than any thing short of mob vengeance or the operation of the criminal laws.

THE ONLY SAFEGUARD.

We wish it distinctly understood here, by every one who desires information on this subject, that every female who practices any of the various operations, more or less disgusting, to produce abortion, does so *at the risk of her health, if not of her life.* There are no *safe* means of abortion. If accomplished, it is a crime; to attempt it is criminal in effect and intent. So think the best men in and out of the medical profession; so say all moralists who have considered the subject.

But there is an alternative course—if that can be called alternative, which precedes in point of time—which may be resorted to, in many cases, with perfect propriety, and without sin in the sight of God and man. Dr. H. R. Storer, in his admirable treatise of " Why Not," copied on page 169, *ante*, says: " Woman's only safeguard is either to restrict approach to a portion of the menstrual interval, or to refrain from it altogether." The latter remedy, " total abstinence," is more easily suggested than practiced. It is not very consistent with animal nature in any of the species, and hardly less so with human nature. In effect, it abolishes the marriage relation; as it interdicts and prevents one of its most practical, pleasant, and vital incidents. It is simply impossible to perpetuate the race, if any great proportion of the people would afflict themselves to this extent.

But the other remedy is practical and available, and commends itself to the good sense, and morality, and prudence of both sexes. Self-denial for a limited time, or during a stated time not too long, is such a safeguard as is consistent with the laws of human health, of reproduction, of reasonable sensual enjoyment; indeed, with the laws of man and God alike.

The following, from a medical writer of some force

and science, is to the point: " The ovum usually reaches
the womb from one to two days after the monthly
flow ceases. After being retained a certain time by
a thin membrane, called the decidua, the membrane
loosens and passes out of the body, taking the ovum
along with it. While it remains in the womb it is,
of course, liable to be impregnated by the semen
from the male; but the moment it is expelled no
impregnation can take place until another monthly
flow. Many French females, who have studied this
subject closely and attentively, are enabled to tell
with certainty when the ovum leaves them, and
they avoid contact with the other sex, except during
the interval between its expulsion and their next
monthly turn. In this way they avoid child-bearing. The usual healthy time during which the
ovum remains in the womb, is fourteen days. In
some females it remains as long as sixteen or seventeen days. Cases of supposed barrenness are frequently those where the *ova* are expelled from the
womb very soon after lodging there. It is, then,
necessary for the husband, if he desires children, to
cohabit with his wife immediately after the menstrual flow ceases. On the contrary, those who
would avoid having offspring, should refrain from
sexual indulgence until the ovum has been expelled,
which is generally the beginning of the third week
after the menstrual flow has ceased."

In the London *Medical Times and Gazette*—one of
the most influential and reliable medical journals in

the world—is the following communication from an
eminent physician: ·

"SIR :—At what, if any, period after menstruation, is conception impossible? Dr. Prosper Lucas, in his admirable work
'*De l'Hérédite Naturelle,*' states, in the most unqualified manner,
that 'there is no day, however distant from menstruation, at
which a woman may not and does not conceive '—vol. ii, p. 917.
Dr. Carpenter, however, in his 'Human Physiology,' is quite
disposed to limit the period of possible aptitude for conception,
and can only adduce one instance, from his own experience, in
which conception followed connection occurring so long as
seven days after menstruation (p. 1004, note, 4th edition).

"Can any of your readers contribute a single case in which conception certainly took place subsequently to the twelfth day after
menstruation had ceased, and not immediately prior to its return?   Are there any works or contributions to the journals in
which this important question is discussed, in addition to those
referred to by Dr. Carpenter (*Op. Cit.*, pp. 998, 1004)?   It is to
be hoped that in the forthcoming edition of his 'Physiology,'
Dr. Carpenter will be able to arrive at a definite conclusion
upon the subject.   I am, etc.,
"Jan. 24, 1863.   A SUBSCRIBER.

"[All such evidence is hearsay ; but on such evidence we know
one case in which conception occurred on the tenth day after
menstruation.—Ed. *Med. Times and Gaz.*] "

This testimony is of the very highest character, in
the medical profession—to the effect that by *waiting
until the eighth day* after menstruation ceases, it is
extremely probable that conception and pregnancy
will be avoided.   If the couple can restrain their
desires thus long—in any state of case where they have
persuaded themselves that it is important or very desirable to prolong the interval between childbearing,

or to avert it altogether—they will find that absence strengthens love, and the embrace is all the more fervently enjoyed because of the removal of danger. Thus a *little* continence—certainly not unreasonable, if a really valuable end is to be subserved—will be repaid by a very grateful enhancement of marital delights.

But we must not be understood as pronouncing this rule *infallible.* On pages 84–5, *ante*, will be found a carefully prepared table of thirty-three cases, of which thirteen are *exceptions to this rule*—*i. e.*, thirteen of them are cases in which, so far as such hearsay evidence can be relied upon, conception took place *after the eighth day.* A number of able statisticians have gathered such results as they believed to be fairly reliable, and have concluded that the law above stated is so positive and universal, that the exceptions are not more than six or seven per cent., *i. e.*, in not more than six or seven out of every hundred cases where the beginning of pregnancy was *ascertained*, did conception take place until after the eighth day succeeding the close of menstruation. M. Raciborski, in his celebrated work "*Sur la Ponte des Mammifères*," has shown that the general rule is, that women become impregnated *immediately before or after*, and even during, menstruation; and the exceptions to the rule were as above stated.

It may be of interest to some doubting ones to state that the writer is personally acquainted with a handsome, healthy, and amorous couple, who—with too

limited pecuniary resources as the satisfactory reason—
have systematically practiced the above rule for over
nine years, with the result of only one child, and that
when they had agreed to waive the rule and add one
to the family. It is proper to say, however, that out
of abundant caution, they extended the time usually
to fourteen or sixteen days. Another couple known
to the writer, successfully relied upon the eight-day
rule, for over fifteen years, when the husband died.

Since the foregoing was in type, in some additional
researches which were necessary, the following from
the able work of Prof. Chas. D. Meigs on Obstetrics,
3d edition, page 151, attracted our attention. On
the reputation of Prof. M., although differing materially from the examinations and conclusions of other
influential medical men, we commend it to the careful consideration of all readers who may be *interested:*

" The modern student of medicine knows that ovulation being a spontaneous act of ovipositing, fecundation of the egg can take effect only after it has been
set at large by the act of the absorbents. Therefore,
he also knows that a woman can be subject to fecundation only within some unknown but short season
succeeding her menstruation. M. Pouchet, of Rouen,
whose learned researches have thrown a flood of light
on this obscure department of physiology, seems to
feel quite sure that the period after a menstruation
during which the discharged ovule remains subject
to fecundation, does not extend beyond the twelfth

day ; and I can not gainsay that decision of the learned naturalist and physician. I am, however, sure that it does extend to the eighth day, and probably beyond it—because Jewish women among my patients, and very prolific ones too, have assured me that they never did violate what they regard as a religious duty, that commands them to avoid the approach of their husbands until eight complete days have elapsed after the entire cessation of the menstrual flux. They have assured me that, in a very large Jewish sect, this law is scrupulously fulfilled; and if so, we have in this religious custom a positive proof that the escaped egg remains apt for fecundation during the eight Jewish days of abstention. Why may it not also be true that M. Pouchet's law of twelve days was correctly ascertained ? There is surely no reason to suppose that the discharged ovule must necessarily perish or decay sooner than the twelfth day after its separation, since it may lie perdu within the fimbria or in the canal of the Fallopian oviduct, ready at any time to meet the conflict with the male zoosperm, which is fecundation."

CHAPTER XVII.

CONTROL OR GOVERNMENT OF THE SEXES.

THIS subject has for years enlisted the attention of many inquiring minds, and furnished the theme of many essays in scientific, medical, and agricultural works. The opinions and suggestions thus put forth are as curious and interesting as they are novel, and some of them ridiculous. Before stating and elaborating the theory which our own observation, during more than thirty years, has convinced us is the true one, we think it due to the labors of investigators to give all the theories—plausible or otherwise—which we have met with.

The *American Agriculturist*—a monthly as venerable for its age as it is notable for its enterprise—in the number for June, 1842, page 63, has the following:

TO BREED MALES OR FEMALES.

We have heard it asserted by distinguished breeders, that in seven cases out of ten, when the progeny was single, they could produce a male or female offspring as they might wish. We have had some experience in this matter ourselves, but never obtained any thing like the certain results above expressed. But we will give the rule:

Should the sex required be represented in the intended parents by the most stout and robust, or the one possessing the highest condition, the rule would require them to be coupled in this condition, increasing their relative diversity perhaps for the occasion, by exhaustion on one side, and augmenting the vigor on the other. If the weakest or lowest in condition possess the sex sought, the physical superiority of the most athletic should be temporarily changed, so far as possible for the occasion, by partial exhaustion or fatigue, as might be done in a variety of ways, especially, if in the male, by sexual exertion.

Some learned writers have contended, that the sex of human progeny is determined by the advanced age of the parent, the one decidedly (*relatively*) the oldest controlling it ; and though true enough frequently, may yet be considered an accidental result. Walker, who has written latest and fullest on this subject, we never considered as any reliable authority; and we have perused his works more for the purpose of noticing the development of his startling theory, than any established or well authenticated principles. The matter is, however, worthy of investigation; and the world will be glad of the experience of such gentlemen as have made sufficiently extended and accurate observations from which to draw general and incontrovertible conclusions.

Somewhat in the same line of suggestion, but more confidently and boldly discussing the relative influence of parents and parent animals on the sex of the offspring, are the following curious observations from an English journal, the Ten Towns *Messenger*, June, 1843 :

The influence exerted by the relative age of the parents in determining the sex of the offspring, I think I shall show to be considerable—all other things, as health and condition, nature of keep, etc., being equal.

If the male is younger than the female, or if they are of the same age, the offspring will probably be female.

If the female is but very little older—a few months or a few years, according to the longevity of the kind of animal—the sex will be doubtful, and probably depend on their relative health and strength at the time of impregnation.

And lastly, if the male be considerably older than the female, while yet his animal powers are undiminished in vigor, the greater the difference, the more likely will it be that the offspring shall be male.

The following table is illustrative of the relative influence of the age of the parents on the sex of the offspring. This table is drawn up from the records of the British peerage; where, of course, every particular of marriages and births has been for ages recorded :

When the husbands were younger than the wives, to 100 girls were born 86 boys.

Where the husbands were of the same age as the wives, to 100 girls were born 94 boys.

Where the husbands were older, from 1 to 6 years, to 100 girls were born 103 boys.

Where the husbands were older, from 6 to 11 years, to 100 girls were born 126 boys.

Where the husbands were older, from 11 to 16 years, to 100 girls were born 147 boys.

It will at once be seen that the influence shown by this table is too striking to be the result of chance. It is drawn up from the ages alone, without taking into consideration any secondary causes; and yet, notwithstanding this, the probability is shown to be nearly as high as *three* to *two* in the extreme. Now, should the analogy hold good between man and domestic animals (and there is every reason to believe it does, in a great measure, with such as produce one or rarely more at a birth), I think it will be granted me, that this influence is sufficiently great to demand our attention.

That the relative condition of the health and strength of the parent animals at the time of impregnation should have some considerable influence in determining the sex of the offspring, where the age and other circumstances are equal, it is easy to conceive, but very difficult to prove. I have no facts to offer on this head; but the very marked manner in which the offspring, in other respects, sometimes takes after one parent, sometimes after the other, successively, is strong presumptive evidence that such would be the case with reference to the sex.

In the United States Agricultural Report, for 1860, issued from the Patent Office, is copied the following article, a translation from the French. It was communicated to the *Journal d'Agriculture Practique*, by M. Martégoute, and furnishes information of value and interest to those engaged in the breeding of sheep :

ON THE PRODUCTION OF SEXES AMONG SHEEP.

The interesting researches of Giron de Bazareingues into generation, and particularly on the production of the sexes among domestic animals, are now known but by very few persons, having the misfortune to be of too remote a date. On the other hand, meeting with a varied reception on their appearance, they have had the fate of all contested things—they have left in the mind nothing but ideas undecided as to their value. Zootechny, in fact, was too little advanced at that period for the art of animal production to think of extracting, from such a study, facts for its use.

Daily observations, conducted and arranged with the calculation in hand, in a sheepfold of great importance—that of the Dishley-Mauchamp merinos, of M. J. M. Viallet, at Blanc, in the commune of Gailhac-Toulza (Haute-Garonne)—have enabled me to comprehend the laws which, according to M. Giron de Bazareingues, preside over the production of the sexes.

The general law which Giron de Bazareingues recognized on the subject of the procreation of the sexes is as follows: The sex of the product would depend on the greater or less relative vigor of the individuals coupled. In many experiments, purposely made, he has obtained from the ewes more males than females, by coupling very strong rams with ewes either too young or too aged, or badly fed; and more females than males by an inverse action in the choice of the ewes and rams he put together.

This law has developed itself regularly enough at the sheep-fold of Blanc, in all cases in which circumstances of different vigor between the rams and ewes have been observed in coupling them. Witness two striking examples of it:

In 1853, births, the issue of young ewes by a Dishley-Mauchamp merino ram, extremely vigorous and highly fed, produced twenty-five males, and nine females only, or 71.73 per cent. of males, and 28.27 per cent. of females.

At a later period the same ram, still in full vigor, having been put to some ewes that had done nursing their lambs—a period at which the ewe is found very weak--there resulted, in 1853, eight male births against four females; and, in 1854, under similar circumstances, seventeen male against nine female births. The two occasions united yielded 65.78 per cent. of males, and 34.22 per cent. of females.

But the following fact has nothing in common with those related by Giron de Bazareingues, and which has been repeated, with small variation, every year from 1853, the period at which the observations I have noted down began.

This fact consists, 1st. In that, at the commencement of the rutting season, when the ram is in his full vigor, he procreated more males than females.

2d. When, some days after, the ewes coming in heat and in great numbers at once, the ram was weakened by a more frequent renewal of the exertion, the procreation of females took the lead.

3d. The period of excessive exertion having passed, and the number of ewes in heat being diminished, the ram also found less weakened, the procreation of males in majority again commenced.

In order to show that the cause of such a result is isolated from all other influences of a nature to be confounded with it, I shall take the years 1855–'56, in which, by the effect of a de-

gree of equilibrium of age and vigor between the rams and
ewes, the male and female births were found, relatively with
each other, nearly upon a par in numbers, being 25 males to 23
females.

The following table, drawn up with the dates of birth, exhibits the facts in detail. The letter M indicates the male and
F the female births.

It will be seen that, the list of births having been divided
into three successive series, and in mean proportions almost
equal, we have for the first, of eleven days, from the 27th December to the 8th January, 13 males against 4 females; for the
second, of nine days, from the 9th to the 18th January, 3 males
only against 15 females; and for the third, of eleven days,
from 19th to 29th January inclusive, 9 males against 4 females:

*Table of the Dishley-Mauchamp Merino Lambing, at the sheepfold
of Blanc, in December and January, 1855–'56.*

FIRST SERIES.

December 27......M.	January 4......M.	January 6......M.
30......M.	4......M.	7......F.
31......M.	4......M.	8......M.
January 3......M.	5......M.	8......M.
3......F.	5......M.	8......F.
3......F.	6......M.	

Males, 76.8 per cent.; females, 23.9 per cent.

SECOND SERIES.

January 9......F.	January 13......F.	January 16......F.
9......F.	15......F	16......F.
11......M.	15......F.	16......F.
12......F.	15......M.	17......F.
12......F.	16......F.	18......M.
13......F.	16......F.	18......F.

Males, 16.66 per cent.; females, 83.24 per cent.

THIRD SERIES.

January 19......M.	January 20......M.	January 24......M.
19......M.	20......F.	24......M.
19......F.	22......F.	29......M.
19......F.	22......M.	
20......M.	23......M.	

Males, 69.23 per cent.; females, 30.77 per cent.

18

At the end of each month all the animals of the Blanc sheep-fold are weighed separately; and, thanks to these monthly weighings, we have drawn up several tables, from which are seen the diminution or increase in weight of the different animals, classed in various points of view, whether according to age, sex, or the object for which they were intended.

Two of these tables have been appropriated to bearing ewes, one to those which have borne and nursed males, and the other to those that have borne and nursed females. The abstract results of these two tables have furnished two remarkable facts:

1st. .The ewes that have produced the female lambs are, on an average, of a weight superior to those that produced the males, and they evidently lose more in weight than these last during the suckling period.

2d. The ewes that produce males weigh less, and do not lose in nursing so much as the others.

If the indications given by these facts come to be confirmed by experiment sufficiently repeated, two new laws will be placed by the side of that which Giron de Bazareingues has determined by his observations and experiments.

On the one hand, as, at liberty or in the savage state, it is a general rule that the predominance in acts of generation belongs to the strongest males to the exclusion of the weak, and as such a predominance is favorable to the procreation of the male sex, it would follow that the number of males would tend to surpass incessantly that of the females, among whom no want of energy or power would turn aside from generation, and the species would find in it a fatal obstacle to its reproduction. But, on the other hand, if it was true that the strongest females, and the best nurses among them, produce females rather than males, nature would thus oppose a contrary law, which would establish the equilibrium, and, by an admirable harmony, would secure the perfection and preservation of the species by confiding the reproduction of either sex to the most perfect type of each respectively.

ON THE ORIGIN AND DISTRIBUTION OF SPECIES IN PLANTS.

Dr. Hooker, of England, in a recently published work on the "Botany of the Antarctic Voyage," in discussing the relations and distribution of species in plants, lays down the following propositions as axioms:

" 1. That all the individuals of a species have proceeded from one parent (or pair), and that they retain their distinctive (specific) characters. 2. That species vary more than is generally admitted to be the case. 3. That they are also much more widely distributed than is usually supposed. 4. That their distribution has been effected by natural causes; but that these are not necessarily the same as those to which they are now exposed.

" Hybridization has been supposed by many to be an important element in confusing and making species. Nature, however, seems effectually to have guarded against its extensive operation and its effects in a natural state, and, as a general rule, the genera most easily hybridized in gardens are not those in which the species present the greatest difficulties.

" With regard to the facility with which hybrids are produced, the prevalent ideas on the subject are extremely erroneous. Gärtner, the most recent and careful experimenter, who appears to have pursued his inquiries in a truly philosophical spirit, says that 10,000 experiments upon 700 species, produced only 250 true hybrids.

" It would have been most interesting had he added how many of these produced seed, and how many of the latter were fertile, and for how many generations they were propagated.

" The most satisfactory proof we can adduce of hybridization being powerless as an agent in producing species (however much it may combine them), are the facts that no hybrid has ever afforded a character foreign to that of its parents, and that hybrids are generally constitutionally weak and almost invariably barren. Unisexual trees must offer many facilities for the natural production of hybrids, which, nevertheless, have never been proved to occur, nor are such trees more variable than hermaphrodite ones."

THE SEX OF EGGS.

Much speculation has been excited or revived as to the sex of eggs, especially at each recurring "*morus multicaulis*" fever or epidemic among the dealers in fancy poultry. In the *Northern Farmer*, October, 1852, page 152, a correspondent contends that there is no difficulty in perceiving the incubation end of a hen's egg, and adds: "Some people can tell a pullet from a rooster; the marke for a rooster is crosswise, and for a pullet lengthwise," etc.

M. Genin, an enthusiastic French agriculturist, wrote a very interesting address to the *Academie des Sciences* on this subject. He says he is able, after three years study, to state with assurance that all eggs containing the germ of males have wrinkles on their smaller ends, while female eggs are equally smooth at both extremities.

In the *Scientific American*, October 3, 1863, is recorded what Charles H. Grower, of Long Island, says in reference to M. Genin's plan for determining the sex of eggs ; that, wishing to have a number of cocks, he put a dozen eggs with rough ends under a hen ; and two males and seven females were hatched out. As to the other part of Genin's plan—if the air bubble is in the center of the end of the egg, a male bird would be produced, or if slightly at one side, the egg

would give a female—he tried fifteen eggs selected as
male, and the result was seven males and six females.

PROCREATION OF THE SEXES OF FOWLS AND CATTLE AT WILL.

In the London *Lancet*, of June 3, 1865—one of
the leading medical journals of the world—is trans-
lated an outline of an interesting memoir presented
by M. Coste, at a recent meeting of the Academy of
Sciences. The learned Professor of Embryology at
the Collége de France, who has already produced so
many remarkable and well known works on this important branch of the natural sciences, related a series
of experiments which he had been led to undertake
by the publication of M. Thury's (of Geneva) theory
on the procreation of the sexes at will. This theory
is grounded on the idea that the ovum passes through
two distinct phases: a first period, which corresponds
with a lesser degree of maturation, and during which
the sex is female; and a second period, one of more
advanced maturation, during which the egg is male;
and according as fecundation takes place at the beginning or at the end of maturation the product will
be male or female.

M. Coste's first experiments were made on mul-
tiparous animals, chiefly fowls. If we suppose the
seminal fluid to be diffused at a given moment on
the ovary, and consequently on the ovules which still
adhere to it, according to M. Thury's theory the first

ovule which tears its covering and issues into the oviduct is necessarily the one which bears the highest degree of maturation. As a single coition permits a hen to lay during twenty days after, it follows that the last ovules which quit the ovary are at the moment of fecundation far less advanced in maturation than the former; these, then, should give male products, and the others female. According to M. Coste this is not the case, and in all his experiments made on hens the first egg was always infertile, while the following were indiscriminately male or female. Fecundation may be supposed to take place in the Fallopian tube, and not in the ovary, the sperm being arrested in the pavilion, or at the extremity of the oviduct. But then, each egg which comes into contact with the fertilizing liquid having attained the same degree of maturation, all the products should be male, which is not the case.

Other experiments were made on rabbits, to which M. Thury's theory was equally applicable; that is to say, females fecundated when the period of rutting has just commenced should give birth to females, and vice versâ. Experience has given a contrary result, and a she rabbit, having been fecundated at the very outset of the period of rutting, produced a greater number of males than females.

M. Coste proposes to examine in a future paper the same phenomena among uniparous animals.

From the American *Homœopathist*, June, 1868,

we copy the following additional in regard to M. Thury's theory :

"The laws of procreation are very little understood, and such are the difficulties surrounding the investigation of this subject, that we may not reasonably hope for a speedy or perfect solution of their varied phenomena. This, however, is an age of bold inventions and heroic devotion to occult subjects ; and we have great confidence that the time is not far distant when *some* of the so-called inexplicable phenomena of our existence will be understood. Something has already been accomplished. No one who has given the subject a passing thought, can doubt that the production of the sexes is governed by laws which are immutable in their operation, and is not, as the common mind is apt to suppose, left to blind chance. The laws which govern conception are a sealed volume to us, or, at least, very little is definitely known; and still less is known of the laws controlling and determining the sex of the human offspring. Among those who have given the subject most thought is M. Thury, Professor in the Academy at Geneva. He observed that the queen bee lays female eggs at first, and male eggs afterward ; that hens produce females from the eggs of their first laying and males from the later products; and so on through the list. Hence, he lays down the following law for the guidance of stock-raisers.

"'If you wish to produce females, give the male at the first signs of heat ; if you wish to produce males, give him at the end of the heat.' This has been tried in several instances, and experiments seem to verify the correctness of the law. Thus the testimony of an extensive stock-raiser and breeder in Sweden is as follows :

"'In the first place, on twenty-two successive occasions I desired to have heifers. My cows were of Schwitz breed and my bull a pure Durham. I succeeded in these cases. Having bought a pure Durham, it was very important for me to have a bull to supersede the one I had bought at great expense, without leaving to chance the production of a male. So I followed accordingly the prescription of Professor Thury, and the success has proved once more the truth of the law. I have obtained from my Durham bull six male bulls (Schwitz Durham cross) for field work ; and having chosen cows of the same color and height, I obtained perfect matches of oxen. My herd amounted to forty cows of every age. In short, I have made in all twenty-nine experiments after the new method, and in every one I succeeded in the production of what I was looking for—male or female. I had but one single failure. All the experiments have been made by myself, without any other person's intervention; consequently I do declare that I do consider as real and perfectly certain the method of Prof. Thury.' M. Thury submitted his plan to the Academy of Science at Paris. Upon their recommendation it was tried upon the Emperor's farm, and, it is affirmed, with complete success.

"Many other experiments have been awarded with like results. These have not been confined to the medical profession, but have embraced many non-professional men, who, as a matter of science or profit, have become interested."

The Hon. William McCombie, in an able and thoroughly scientific article on the "Breeding and Care of Cattle," in the Ohio Agricultural Report for 1867, says:

"It is well known to breeders of cattle, and, I believe, of sheep, that there are particular races that are celebrated, and upon which you can calculate that they will never propagate an inferior animal. Specimens not so desirable will now and again appear, but the blood in these and the divergence will not be great from the desired type. Again, there will be one race noted for producing celebrated males, and another for producing celebrated females. A bull may be introduced that is a great getter of bull calves, yet the change may not be to the advantage of . the owner, as the female calves will not be bred of so high an order.

"Professor Thury, of Geneva, has written a very interesting paper on the law of the production of sexes. In a letter to me, dated 14th of February, 1864, he says: 'There are, if the owner pleases, two periods of heatings; the one, the general period, which shows itself in the course of the year, following the seasons; the other, a particular period, which lasts, in cows, from twenty-four to forty-eight hours, and which reveals itself a certain number of times. It is this particular period, lasting from twenty-four 'to forty-eight hours, the commencement of which gives females, while its termination gives males.

" ' In order that we may obtain a certain result, we must not cause the same cow to be covered twice in succession, at an interval too short; for the (generative) substance of the bull preserves itself for a time sufficiently long in the organs of the cow. In the experiments made in Switzerland, we have taken the cow at the first certain signs of heating, for the purpose of obtaining heifers; and at the termination of the heating, for the purpose of obtaining males. The result of these experiments is, that we do not yet know what is the relative length of time which gives females, and the time which gives males. This would form an interesting subject of examination. I am of opinion that various circumstances must be regarded as influencing the relative period, so as to alter the moment of conception, and that the season must exercise considerable influence. In such questions as that which forms the subject of my little work, we physiologists should learn much from men of practice and experience, such as you who have afforded proofs of their knowledge. The best results will follow when the raisers and experimentalists direct their attention to the same object.'

"The experiments conducted in Switzerland were decisive in support of Professor Thury's theory. In a trial of twenty-eight cows, it proved correct in the whole number. In the selection of the male, you will have to consider the faulty or defective points in your cows, with a view to correct them. As far as possible—pedigree being right—you ought to purchase the bull that is strong upon the points where your females are faulty. If this is not duly attended to, the defect or malformation may be aggravated. But, although the bull selected possesses the excellence wanting in the cows, he ought, of course, not to be very deficient in other points, else the cure may be worse than the disease. If possible, he should be taken from a pasture not

superior to your own. Docility of temper in male and female
is indispensable. ¯Inexpressible mischief may be done by the
introduction of wild blood into the herd, for it is sure to be
inherited."

SEX DETERMINED BY THE MOVEMENTS OF THE FŒTAL HEART.

Dr. Frankenhäuser asserts in the *Monatsschrift für
Geburts-K.*, that while paying attention to the comparative frequency of the beats of the fœtal heart,
before and during labor, he noticed that the heart
beat with less rapidity in the male fœtus than in the
female. The pulse in the former is on the average
124, in the latter 144. Upon these data the author
has almost invariably predicted the sex. If the asser-
tion is true, the correct explanation of the fact will
not be very difficult; it is known that as a general
rule the pulse of girls is more rapid than that of
boys of the same age.

Dr. Steinbach, while assistant physician to the
Lying-in Hospital of Jena, in the summer of 1859,
carried on his observations upon fifty-six pregnant
women—by auscultations every morning and after-
noon, from the day after their admission until labor
set in. From these pulsations of the fœtal heart he
determined the sex of the child forty-three times out
of the fifty-six. He counted the beats of the fœtal
heart during a quarter of a minute—finding the
mean number for boys to be 131, and for girls 144;

the extreme numbers for boys being 124 and 132, and for girls 133 and 147. He confirmed the observations of others—that there is not a diminution of the pregnancy of the fœtal pulse corresponding with the increasing age and development of the fœtus.

Dr. James Cumming communicated to the Obstetrical Society of Edinburgh, in May, 1870, some interesting investigations on this subject:

TABLE I.—MALES.

" The first case was one of twins, the heart of the one fœtus was heard in the right groin beating 110 in the minute, and on delivery it proved to be a male; the second heart was heard in the left hypochondrium, beating 154, and on delivery it was found to be a female.

	Per minute.			Per minute.
2. Fœtal pulsation,	138	15. Fœtal pulsation,	116
3. " "	138	16. " "	120
4. " "	135	17. " "	120
5. " "	130	18. " "	138
6. " "	130	19. " "	125
7. " "	132	20. " "	140
8. " "	132	21. " "	140
9. " "	140	22. " "	137
10. " "	132	23. " "	140
11. " "	140	24. " "	141
12. " "	136	25. " "	122
13. " "	133	26. " "	120
14. " "	134		

TABLE II.—FEMALES.

	Per minute.			Per minute.
1. Fœtal pulsation,	150	9. Fœtal pulsation,		140
2. " "	142	10. " "		152
3. " "	140	11. " "		140
4. " "	150	12. " "		143
5. " "	144	·13. " "		144
6. " "	140	14. " "		141
7. " "	140	15. " "		160
8. " "	144			

"From these two tables it seems that when the pulsation varies from 120 to 140, the probability is that the fœtus will be a male, and when the pulsation varies from 140 to 160, the fœtus will likely be found to be a female. But there are some exceptions to these facts. In three cases in which the pulsation was from 150 to 160, the fœtus proved to be a male ; and in fifteen cases in which the pulsation varied from 116 to 138, the fœtuses were found to be females. It, therefore, appears that there is less frequent variation in the pulsation in the male fœtus than in the female; or rather that there are fewer cases in which the heart's action exceeds 140 in the male, than that it falls below that number in the female."

SEX DETERMINED BY THE MORNING SICKNESS, FORM, AND APPETITE.

A French medical writer of considerable celebrity attacks, as a *popular error*, the opinion that there is a mode by which male or female offspring may be produced at will. And yet he concedes the truth of one great point claimed by some of these errorists, viz.: that some observing persons can quite certainly tell the sex of the child a month or several months before its birth. He argues thus:

"No consequence whose theory of the mysteries of reproduction is correct, they are agreed on certain points, which shows this to be impossible. There are tolerably conclusive rules, however, for telling the sexes of children before they are born, and were I to be guided entirely by the testimony of my own experience, I would say that these rules were infallible. Ladies experience more sickness with boys than with girls, probably because they are generally larger and more lively. Their foreign appetites are also of a stronger, better defined, and more natural character. For instance, with the one they will long for meat, spirituous liquors, etc., with the other for chalk, isinglass, and various substances, which would be quite repugnant to them at other times. Again, roundness of form promises a boy, whereas when the tendency is nearly all to the front, and the hips and back give but little evidence of the lady's situation, the great probability is that the little stranger is a girl. At all events these indications never deceive me. Old women say that boys lean to the right side and girls to the left; also, that boys improve the beauty of the mother's countenance, while girls detract from it; notwithstanding the latter cause them the least sickness. But these signs I hold to be mere fables, as I never could see their philosophy established. On the whole, when a lady in the family way is prone to sickness in the morning—longs for food of an invigorating quality—and carries her increase of form rather all round her than in any particular place, the chances are altogether in favor of a boy—perhaps, indeed, imperatively so; whereas, if her symptoms are otherwise, and as described above, she will in all probability be delivered of a girl. I claim to be original in these discoveries, as I am not aware that any other physician has studied or written on the subject."

SEX DETERMINED BY THE PERIOD OF CONCEPTION.

A farmer in St. Louis County, Mo., writes to the *Field and Farm* newspaper that he has already been able to predict with certainty, in many cases, the sex of the future infant. More than thirty times among his friends he predicted the sex of the child before the birth, and the event proved nearly every time that he was right. The practical lesson is simply this: If a husband desires male children, let him not expose his wife to conception during the early part of her intra-menstrual period, and *vice versa.*

SEX CONTROLLED BY THE DATE OF IMPREGNATION.

A medical writer of some ingenuity and enterprise, but of no great influence in the profession, has struck out an idea about controlling the sex, which is more original and novel than substantial. Proceeding upon what we claim to be the correct basis—that the ovaries themselves are sexual, the one giving off male ova only, and the other female only—he insists that there is an unvarying alternation of male and female ova developed and discharged from the ovaries in the process of menstruation; and that the wife has only to follow certain definite rules which he lays down— but which are difficult to comprehend, and still more difficult, if not impossible, to *observe* with any certainty—in order to fix the knowledge of the sex of

her child. Supposing it were as simple as it really
is complicated, and as reliable as the inventor per-
suades himself it is, it does not claim to have a starting point until after the birth of the first child or a
premature labor at so late a period as to leave no
doubt of the sex of the fœtus. Such a weakness in
his theory would be fatal to its practical value in the
thousands of cases, like that of Louis Napoleon, the
recent Emperor of France, whose Empress had borne
him but *one* heir apparent to the throne and to his
individual wealth. It is the *first* child about which
many parents are most deeply concerned, for bad
health, poverty, unwillingness to endure the sufferings of pregnancy and childbearing, and the annoy-
ances of nursing, and other *controlling* reasons, very
often limit the number of "olive plants" to the one
which introduced the young wife to the pains and
penalties of maternity and all its trying responsi-
bilities.

OLD WOMEN'S NOTIONS ABOUT CONTROLLING THE SEX.

These are so numerous, and often so silly and unreasonable, that if we could here aggregate and col-
late them all, it would furnish an amusing but not a
very instructive chapter.

Some stoutly maintain that if impregnation takes
place while the moon is increasing, and before its
full, the progeny will be a male; but if in the wane
of the moon, then a female.

Others as positively insist that the will of the parents controls the sex—they to *wish* heartily for the sex preferred, while in the act of coition, and to picture in their mind's eye the image and sex of the one longed for.

Others contend that the difference in the ages of the parents control the sex; that if the father be older and stronger, boys will preponderate in number, and *vice versa.* But any body, upon a little careful thought as to the history of their neighbors' households, will readily see how futile is this theory. As we write, the following recur to us: The father, thirteen years older than his wife, three boys and four girls; the father, six years the older, five boys and six girls; the mother, nearly two years the elder, six children, three of either sex; the mother, three years the older, five boys and no girls; the mother about thirteen years the elder, only one child, a boy; the father, fifteen years the elder, only two children, girls; the father, nearly sixteen years the elder, five healthy and strong girls, and one puny boy.

Other old women require the mother, while sleeping, to face particular points of the compass, according to the sex desired.

Others assure listening married women that to accomplish a change of the sex of their offspring they must "turn the bed around," which easy arrangement, they say, is sure, if persevered in.

Another of the notions entertained by the large class of people who wish to be thought wise upon this subject is, that the attitude of the husband and wife, in coition, controls the sex, *i. e.*, the approach of the husband must be on a particular side of the wife, or the pressure of his body be greater upon one side, to insure a male, and the reverse for a female.

OUR METHOD FOR CONTROLLING THE SEX.

We have now given all the different theories for controlling the sex—obtained in an extensive and systematic course of examination, continued during more than twenty years—including the methods suggested by the scientific as well as the unlearned, by the veterinarian and stockbreeder of moderate pretensions alike with the most nicely-investigating physiologist in all the most enlightened French, German, and English-speaking countries. We have not even omitted some of the popular and simple old women's theories, giving all for just what they are worth. It may be that our painstaking, in this direction, will help to make a progress greater in the next quarter of a century than in the last twenty centuries.

We must now state our theory—not ours in the sense of discovery, for it was discovered nearly two thousand years ago—but ours in the sense that it has been a favorite with us since the year 1837, and that we think we are presenting to the public some valuable, if not incontestible proofs of its correctness, and helping to give it a prominence it has never re-

19

ceived, but which it most assuredly deserves. That it will henceforth attract the attention of the medical profession to an extent not now realized, we do not doubt; and this publication give an impetus to the scientific and practical investigation of the whole subject of the reproduction of the sexes.

ILLUSTRATIONS FROM CASES OF EXTIRPATED OVARY.

The theory we advocate is simply this: The testes and the ovaries—the distinguishing features of the male and the female organs of generation—are not only dual but sexual; they are essentially, and in fact, male and female. And if there be any fact yet developed, in investigations of physiological subjects akin to this—any extraordinary result of any nice and dangerous surgical operations involving the health or soundness of the ovaries and circumjacent organs—which can not be explained consistently with the idea of a male and female ovary, and a male and female testicle, we have simply to remind the observing and the doubting reader that there are, in actual life, many cases of hermaphrodites, real or supposed, in which the sexes are so mysteriously blended as to render it impossible, except *post mortem*, to ascertain their rightful position. With even greater satisfaction, we have to remind them that the real history—the life, movements, and disappearance, and the several times and seasons and developments

—of the human ova are still practically unknown, although theorized about and guessed at by ten thousand writers and investigators.

We claim, 1st. That the right testicle contains the semen which produces the male, and the left testicle the semen which produces the female.

2d. The right ovary of the female is masculine, while the left is feminine. From which it follows that,

3d. If a man be deprived of the left testicle, he can beget only boys, and if of the right, then girls alone will be the product; and

4th. If a woman lose her right ovary, by surgical operation or by disease, she can bear girls only, while if the left one be destroyed boys only will be the result.

Any variation or departure from these rules, if any be found, we claim further, will be exceptional and anomalous; and will be as far from disproving, or unsettling, well-founded faith in the rule just laid down, as the two cases of double ovariotomy in the hands of Dr. H. R. Storer, of Boston, and the third case of the same in the hands of Professor Edward W. Jenks, of Detroit, Michigan, followed in each by a sanguineous discharge, rather scanty in amount, but regular, would be from subverting the old idea of such intimate connection between the ovaries and menstruation, as to make the latter inseparable from and dependent upon the former. Such exceptions or anomalies are incident to all theories of the laws of

physiology and physics; and while they do not sub-
vert any established rule, but only help to confirm it
—as expressed by the old proverb, "the exception
proves the rule"—we do not hesitate to say boldly
that it will require many and well-grounded exceptions to disprove the correctness of the rule we adopt.

And right in this connection, we deem it best to
give a few illustrations from the recorded cases of
some of the ablest surgeons in ovarian diseases, in
both England and the United States.

One of the earliest successful cases of ovariotomy,
where the subsequent history of the patient was followed to the point which shows its bearing upon the
theory we advocate, was that of Anne Jones, aged
25, four years married, and the mother of three
children—in *Med.-Chir. Trans.*, xxxii, p. 64. In
November, 1846, Mr. H. E. Burd, surgeon to the
Salop Infirmary, removed from her a large tumor,
involving with it the left ovary. The operation,
though unknown when it was begun, was when the
patient was three months advanced in pregnancy,
and of course cost the life of the fœtus. The recovery was good, and on April 4, 1848, she gave birth
to a fine, strong male child.

We have already mentioned, *ante* p. 193, from the
Cincinnati *Lancet*, 1852, p. 182, a case where Dr.
———— extirpated, by a peritoneal section of nine
inches, from Fanny Gould, an ovarian cyst consisting
of hypertrophy of the left ovary, etc. During the
winter after, the catamenia appeared regularly, and

in April, 1850, she married. In January, 1851, her menses ceased; and 282 days after, on October 9, 1851, she was safely delivered of a male child, weighing seven pounds.

In 1855, Mrs. H., of Brown County, Ohio, was operated on for an ovarian tumor, losing with it the left ovary. Her recovery was rapid and fine, and she soon resumed her household duties. Three years later, when visited by the surgeon, at the request of a friend, he found that in the interim she had given birth to two children, both boys, alive and hearty.

Mrs. H., of Preble County, Ohio, in 1864, in an operation for ovarian tumor, was deprived of the right ovary. She has since given birth to a girl.

Mrs. P., of Darke County, Ohio, not long after marriage, in August, 1863, found herself slowly and steadily increasing in size—as all supposed, the usual desirable sequel to the new state. But as there was no quickening about the time she had been told to expect it, and as the growth had become about stationary in size, and seemed increasing in hardness, medical aid was invited, and a diagnosis developed ovarian tumor. In time, she was operated upon successfully, losing the left ovary. Thirteen months after, and at full time, she gave birth to a male child, still-born. In one year after, she gave birth to twins—*both females*, still living. On the 31st of January, 1870, she again became a mother—this time the right sex, a boy.

We are thus particular in mentioning this last case,

because of its seeming departure from the law of sex which we advocate. We are not satisfied, however, that it is an exception, and have never yet been able to trace up any other case like it. All others substantiate the rule—this one alone presents a difficulty. We are convinced that some portion of the left ovary, or some ovules belonging to it, were overlooked and not removed with the tumor; just as in Mr. Longstaff's celebrated case of medullary sarcoma, in *Med.-Chir. Trans.*, xviii, p. 255, he says it seems "almost incredible that a person with such extensive disease in various viscera, and with such a disordered state of the general health, could have become pregnant. It would be a difficult point to decide, pathologically or physiologically, how it was possible for impregnation to have been effected, when, as was shown by dissection (*post mortem*), all the natural structure of the ovaria had been destroyed by scirrous and fungoid tubera. Perhaps, at the time of impregnation, some portion of the ovaries might have contained a vesicle capable of effecting the purpose, and then the changes produced by utero-gestation excited the vessels of the ovaries in accelerating the morbid growth which affected them. I feel the more inclined to this opinion from having observed (by a daily examination), that the liver, during the three weeks that I attended the patient, underwent a manifest enlargement."

If a similar examination could have been made, at the right time, of the patient, Mrs. P., we are satisfied that it would satisfactorily have cleared up whatever of mystery and difficulty the skeptical reader may insist on evolving from the fact that the twins were girls, when they should have been—in strict consistency with our theory—boys. It should be borne in mind, too, that of all the mysteries of human physiology the ovaries and their ovules, their life and vivication, or destruction, as already suggested, are the most. inexplicable and mysterious. Unnumbered cases of extra-uterine pregnancy, of pregnancy in the Fallopian tubes, of pregnancy where the fœtus was found in the parenchyma of the uterus, and again in the substance of the parietes of the uterus, etc., etc., illustrate the complications of the female generative system.

ILLUSTRATIONS FROM EXPERIMENTS WITH ANIMALS.

A physician of learning and skill, an enthusiastic and patient investigator, some years ago instituted several sets of experiments upon different animals, whose movements could be carefully watched and the results faithfully noted. Those upon dogs were as follows:

Two dogs were castrated in part, each of the left testicle, and after cure were impounded severally with two bitches. In due season one bitch had a litter of seven, and the other of four, all males.

The same dogs were subsequently inclosed with

other bitches, with a like result—five and six males respectively.

Another dog was altered, this time losing the right testicle. As expected, his get proved to be all females.

Two bitches were spayed—from one the right and from the other the left ovary was taken. Inclosed with perfect dogs, the one which had lost the right ovary brought forth six females, while the other had only four, all males.

The bitch from which the left ovary was removed was afterward inclosed with the dog whose right testicle remained; result, seven vigorous dogs, no females.

The other bitch was afterward inclosed with the same dog, and no issue followed—palpably encouraging the theory of the sexes of the ovaries, the semen from the right testicle refusing or failing to impregnate the left ovary.

The writer frequently saw, about twenty-five years ago, a pointer slut which had been entirely spayed, on the supposition that she would cease to come in heat. But, true to her sex, her returns of heat were regular, and all the dogs on the surrounding farms made such constant court to her as to prove a serious annoyance and compel the owner to part with her. She was remarkable for her sleek, round appearance, was very affectionate and playful, but too fat to be as fleet as the males of her grade.

The experiments upon rabbits had the same result as with the dogs. Two rabbits were castrated of the right testis, and after recovery permanently separated from the warren and shut up in a pen with four females. As long as that separation continued every litter of young was composed of females entirely—the male parents being physically hindered from perpetuating their sex. A similar experiment where the left testis was removed, produced the contrary result, the increase being of the he sex.

Similar experiments were attempted upon cats, but they proved too unmanageable, enough of them making their escape from confinement to prevent any reliable results.

Two experiments upon hogs were more successful—one boar being castrated of the right, and the other of the left testis. Being separated and inclosed with young sows, the operator at the full time found eleven sow pigs in one litter, and eight boar pigs in the other.

A friend of the writer informed him of a case with which he was familiar, when quite a young man. Among his father's stock of horses, in keeping a substantial hotel in a country town in the west, were two beautiful matched bays. During the year and a half of his ownership, one of them gave frequent trouble by his disposition, at most unseasonable times and even under quite mortifying surroundings, to leap upon the mares, in the stable or stable-lot, belonging to visiting guests. Curiosity prompted inquiry of his former owner, and it was ascertained that he had been castrated of his right testicle only, the other being out of sight, either wanting altogether or drawn up in the belly. Examination gave proof of the fact, and he was henceforth known as an original, with the left stone perfect, but not visible. As he was the getter of several mare-foals, and of no horse colts, some surprise was excited, and some mysterious suggestions ventured, but the physiological reason was not suspected.

We have read somewhere of a circumstance quite similar—which happened in Florida, while the United States cavalry troops (or dragoons, as they were then styled) were hunting down the miserable remnant of Seminole Indians, who still held out under their brave and eloquent chief, Osceola. Some of the mares belonging to a company at one of the forts, which had been running out on the everglades, were quite unexpectedly found with foal. As it was not known that any other than geldings and a few mares had been purchased for the service, a watch was set, and the cause of the trouble found to be a horse which had one testicle, the left one, the right having been removed. Only female colts were dropped— a fact which the soldiers were never able to account for.

PRACTICAL INSTRUCTIONS.

The reader is now ready with the inquiry, "How do you propose that we avail ourselves of this theory? Admitting that all you claim is natural and reasonable, and satisfactorily established by your reasoning and illustrations, how may we put it to a practical test, and enjoy in our own households the results you promise?" Simply thus: Let the wife who desires female offspring, immediately after connection lie upon her left side; and continue quite regularly upon that side, when recumbent, until satisfied that she is in a family way.* And, so if the

---

Note.—A gentleman, then contemplating matrimony, to whom the writer, in 1863, suggested the above instructions for his guidance in married life—although his own experience since has verified to the letter the rule above—has written out the following *additional* suggestions, which he thinks may be acceptable to some extra-cautious persons :

If a boy be desired, let the husband, during the act of coition, assist the right testicle to discharge, by elevating it with the hand to the upper part of the scrotum, and retard the action of the left testicle by bearing it down. This operation can be performed simply with the thumb and index finger, and does not in any way interfere with the pleasures of the nuptial act. Immediately after coition, let the wife recline on her right side, and stay in that position until the semen has time to find its lodgment in the place assigned for it.

If a daughter be desired, follow exactly the opposite course—*i. e.*, the husband assisting the left testicle to rise, and retarding the action of the right ; and the wife lying on her left side immediately after coition.

The same rule can be made to work with equally satisfactory results in stock-raising. If a stock-breeder desire male offspring, let him watch the operation of the male in coition, and if he sees the right seed ascending in the scrotum he need not interfere; but if the left rise, let him retard its motion by catching hold of it gently, and pulling it back, and at the same time assist the right one to rise by giving it an upward pressure. For females, take directly the opposite course.

wished-for sex be a boy, let her, after coition, systematically and regularly lie upon her right side, until impregnation is fully established. The very natural explanation of such simple directions is this: The semen of the male, although full when healthy of living spermatozoa—just as water is alive with animalculæ—is, like water, a fluid, only much thicker and mucilage-like. As it is the law of all fluids to seek their level, by flowing to any spot or in any direction down-hill or to a lower level, so the semen injected into the wife—if she retain her lying position—naturally flows forward, through the Fallopian tubes, to meet and impregnate the ovule descending or passing out from the ovary which is undermost—i. e., the left ovary, if the female be reclining upon her left side, and the right ovary if she be lying upon her right side. For this reason it is physically unlikely, almost impossible, for impregnation to take place between human beings in a standing position, unless the female very soon after lie down and keep quiet.

These directions have enabled many parents, to whom the information was imparted in the confidence of private conversation, to control the sex of their children—some of whom have grown up to years of discretion, and a few have already entered the marriage state. And, further, they really explain the secret of the old woman's recipe, "Turn the bed around;" because the very fact of turning the bedstead around, in conjunction with the habit already formed of the husband lying upon the out side of the bed, incidentally changes the position of the wife in the bed, and causes her to lie upon the other side from that occupied before the bedstead was moved.

It is an ascertained fact that very few—the wonder is that there should be any—husbands and wives sleep facing each other. Most usually, the husband sleeps upon the outer side of the bed; and the wife, after connection, turns away from, or with her back toward, her husband. Any husband, with a little care, by watching and quietly directing the position of his wife when in bed, can control the sex of his child, unknown to his wife; and the wife, as she is mistress of the situation, need only exercise a little thoughtfulness, to secure the cherished object, and without the husband's knowledge if she chooses.

ILLUSTRATIONS.

For over thirty years the writer had the opportunity of observing a married lady, a neighbor of his, whose first and fourth children were daughters, who died in infancy. Several sons, bright and noble boys, could not fill the aching void in the mother's bosom, and God, in mercy, sent her beautiful twin girls, so remarkably alike that bracelets, with their names engraved, were absolutely necessary to distinguish them. Within a year one was taken, and before three years the other, when the mother's heart rebelled; and in wild agony of spirit, she blamed God for thus bereaving her, and became a raving maniac. In time, her mind was restored ; but for years after, up to the period.when the writer last saw her, she was the subject of a dull and painful melancholy, or was so quick-spoken, exacting, and fault-finding as to make life burdensome to all her household. She gave birth to several sons after the twins ; but the want of the knowledge above, made hers a life of complaining and ingratitude.

One of the most interesting married couple we ever knew were blessed with daughters in their first two children—now blooming in young womanhood, highly cultivated and beautiful: The husband playfully, yet half in earnest, threatened to visit a neighboring orphan asylum and select some promising motherless boy to adopt as his son. The dutiful wife, sympathizing with the longing, but pained at the suggested alternative (she had just learned the *secret* of this book), promised, if he would only wait patiently, to present him with a little son, or, in default, to go with him to the asylum and assist in selecting one for adoption. She made good her promise, and has faithfully tested the practical value of the foregoing instructions—blessing her husband with six loving children, three of either sex.

A distinguished physician in a large western city, who completed his education in Europe, and is now a professor in a prominent medical college, had four sons in succession. A gentleman who had *proved* the virtue of our rule, sounded this physician on the subject, but received no other encouragement than courteous attention. The physician, however, concluded there was no harm in trying it, and in due time his wife presented him with a daughter, their last child. He thus gives silent testimony to the value of that which he would not assent to or approve.

Another physician, less known to fame, but with a more successful and lucrative practice, had three daughters in succession; when, being consulted about the probable reliability of the above secret, promptly said it struck him rather favorably; at any rate, as it cost nothing, and required no sacrifice, and no personal self-denial or inconvenience, he advised him to *try it*, and he himself would do likewise. He is now educating a promising son, with his own profession and a partnership in view; while others younger are growing up.

Another physician, in the suburbs of the second largest city in the west, in conversation with a very intelligent gentleman, in whose family he practiced, had the foregoing secret suggested to him. Quite an animated and rather scientific discussion ensued, the doctor quite ingeniously and warmly opposing, while his friend, with admirable tact and skill, advocated its practicability and importance. The discussion was renewed subsequently, but the doctor refused to be convinced, while the other was evidently more than ever wedded to the theory. In less than ten months after, the doctor called upon his friend, reminding him of the conversation, and told him that, on the day after their last discussion, being called upon by a gentleman whose family olive-plants were only two, and they girls, for professional advice as to how to arrange for a little son next, he promptly assured him there was no rule for it, and he could not advise him. The patient's confidence in the ability and professional skill of his favorite physician was not even staggered at this asseveration, quite earnestly repeated. So he told him he knew, and he *must* tell him; that the peace of his household made it important. The doctor, to get rid of his now troublesome customer, with a quite confidential air, posted him on *our* theory, and was glad to see him go away satisfied. True enough, he called to say that he had just delivered the gentleman's wife of a fine, healthy boy, at full term; and received, as a compliment for having imparted the secret, the handsome present of fifty dollars, independent of his regular obstetrical fee. The gentleman's faith in his family physician has steadily strengthened.

One of the most successful young merchants of Cincinnati, within a few months after his marriage, met, at a fashionable party in a neighboring city, an old friend who inquired after his wife's health. Receiving an equivocal answer, from which he guessed the situation, he playfully remarked that he could tell the sex of the child, if the merchant could give correct replies to several questions. The answers were given, and the friend prophesied him a boy. Several years after, the parties again met in front of the merchant's store, and while conversing, a very handsome little boy came running to his father, with a child's earnest request for the purchase of a toy. The father introduced him to his friend, and presently observed that he was the promised son.

An ambitious young lawyer, in an interior town, was confidentially informed of the foregoing rules on the day of his wedding, and expressed himself anxious for the first child to be a male. The first night of the honey-moon was spent on the Erie Railway, and the position of the berth in the sleeping coach was such as to force the wife to sleep on her left side. Immediately on returning from their extended bridal trip, the husband made known the condition of his wife, and also expressed a fear that, if our rule was correct, their first offspring would be a girl, which fear was realized in due time. Their second child, however, by following the advice given, proved of the sex he so much desired in the first. Fonder or prouder parents it would be difficult to find.

A gentleman, whose first child, by following the foregoing instructions, was a boy, was a little incredulous for twenty months longer, until his wife presented him with a daughter. In the exuberance of his joy, he called at the room of the writer, but not finding him in, left this hasty note upon his table: *"My wife has a fine little girl. Your rule is a success. There is a fortune in it."* We have since seen

20

the happy father, and it would be difficult to express in terms the strength of his faith, it is so confident and enthusiastic.

An honest son of the Emerald Isle, named Dennis C———, had won upon the good graces of a gentleman, by courtesy and fidelity, in the business where he was employed—so that the gentleman one day, learning that he was sparking, playfully told him that when he should get married, he would tell him " how to fix for a boy or a girl either." Dennis was married during the absence of his friend, but on meeting him, two months after, and congratulating him, was told that his wife was already " fixed." He then informed Dennis that his child would be a girl, which, to his serious disappointment, proved true. Following the directions, however, the next girl was a boy, sure enough. Seeing the gentleman passing through the city market, he seized him earnestly by the arm, ejaculating, " I 've got it, I 've got it, but I do n't believe it yet," with the doubting disposition so characteristic of his countrymen.

But enough. Any man who has carefully and inquiringly perused the foregoing pages, and has not absorbed faith enough to be *willing himself to test the virtue of the theory* advocated, deserves to be an Irishman with all of Dennis C———'s lack of credulity, except that he does not deserve to have such pleasant grounds for faith as Dennis had. Or, rather, he ought to be a lineal descendant of the great Syrian captain, Naaman, who, " despising the day of small things," went away very angry because the prophet in Israel would not come out to see him, and, with much form and solemnity, cure him of his leprosy, instead of sending him word to go and bathe in the waters of a river in a foreign land, whose people and their river he regarded with disdain. The above are all *test* cases, by intelligent persons now living, all within the personal knowledge of the writer. And while there is much mystery still about the reproduction of the sexes at will, they know that simpler directions, so easily followed, have never given more positive and gratifying results—bringing joy and rejoicing to their households.

THE END.

GLOSSARY.

Animalcule, a small animal, seen only with the microscope.
Cervix, the neck.
Cellulitis, a wound.
Catamenia, the menses.
Dysmenorrhœa, difficult menstruation.
Dorsal decubitus, the act of lying flat on the back.
Enciente, pregnant.
Epithelial, relating to the epithelium (see below).
Epithelium, the thin layer of epidermis (outer skin), which covers
    or is the lining of such parts as the nipple,
    mucous membranes, lips, pharynx, œsophagus,
    vagina, and entrance of the female urethra.
Epididymis, a small oblong grayish body lying along the superior
    margin of the testicle.
Encephalic, belonging to the head or brain.
Etiology, branch of medical science treating of the cause of
    disease.
Fungoid, from fungus, the mushroom order of plants.
Fistula, chronic abscess.
Fallopian tube, a canal to conduct the sperm to the ovarium,
    and the fecundated ovum back to the uterus.
Fibroid, fibrous, tumorous.
Fœtus, the unborn child.
Hypnotic, a medicine which causes sleep.
Hypertrophy, the state of an organ or part of the body in which,
    from increased nutrition, its bulk is augmented.
Incestuous, guilty of incest.
Lymphatic, distracted, frantic.
Liquor Amnii, the fluid which envelops the fœtus during the
    whole period of utero-gestation.
Leucorrhœa, a discharge of a white yellowish or greenish mucus,
    resulting from inflammation of membrane lin-
    ing the genital organs of the female.
Monorchide, one who has only one testicle.
Metritis, inflammation of the uterus or womb.
Mammæ, the breasts.
Marjorum, an aromatic fragrant plant, much used in cooking.
Medullary, marrow, relating to the marrow, or analogous to
    marrow.
Ovaries, organs of the female in which the new being is generated.
Ovulum, or *ovule*, a small egg.

Ovum, the body formed by the female in which after pregnation the development of the fœtus takes place.

Os tincæ, mouth of the womb.

Os, the mouth.

Phimosis, a condition of the prepuce in which it can not be drawn back so as to cover the glans penis.

Prepuce, prolongation of the foreskin of the penis.

Pelvis, open structure at lower extremity of the body inclosing internal urinary and genital organs.

Phthisis, consumption of the lungs.

Phlebitis, inflammations of a vein.

Placenta, a soft spongy body connected with the fœtus by the umbilical cord.

Peritonitis, the smooth membrane covering the whole internal surface of the abdomen.

Pyæmia, a dangerous disease produced by the mingling of the poisonous matter of pus with the blood.

Polypus, tumor with narrow base resembling a pear.

Puberty, the age at which people are capable of begetting or bearing children.

Quaternions, in four parts—four children at a birth.

Sciatica, rheumatic affection of the hip joint.

Souffle, a sound produced by the passage of air through the bronchial tubes into the air cells.

Sarcoma, fungus, or any species of excrescence having a fleshy consistence.

Spermatozoa, the seed of the male.

Scirrhus, a kind of cancer.

Speculum, a mirror or looking-glass.

Semen, the male generative product.

Sinapism, a poultice or blister.

Tubera, a tumor.

Umbilical cord, a cord-like substance which extends from the placenta to the navel of the fœtus.

Uterus, the womb.

Urethra, the canal through which the urine is conducted from the bladder.

Uterine tube, tube leading to the womb.

Utero Cervix, neck of the womb.

Vaginal Canal, a canal which leads from the external orifice to the uterus or womb.

Vagina, see vaginal canal.

Vesical, pertaining to the bladder.

Viscera, entrails.

INDEX.

www.ingramcontent.com/pod-product-compliance
Lightning Source LLC
Chambersburg PA
CBHW021504210326
41599CB00012B/1132